SpringerBriefs in Electrical and Computer Engineering

More information about this series at http://www.springer.com/series/10059

Zhe Wang • Wei Zhang

Opportunistic Spectrum Sharing in Cognitive Radio Networks

 Springer

Zhe Wang
The University of New South Wales
Sydney
New South Wales
Australia

Wei Zhang
The University of New South Wales
Sydney
New South Wales
Australia

ISSN 2191-8112   ISSN 2191-8120 (electronic)
SpringerBriefs in Electrical and Computer Engineering
ISBN 978-3-319-15541-8   ISBN 978-3-319-15542-5 (eBook)
DOI 10.1007/978-3-319-15542-5

Library of Congress Control Number: 2015933277

Springer Cham Heidelberg New York Dordrecht London

Printed on acid-free paper

Springer is part of Springer Science+Business Media (www.springer.com)

Recommended by Sherman Shen

Preface

Rapid growth of wireless communication services in recent decades has created a huge demand of radio spectrum. Spectrum scarcity and utilization inefficiency limit the development of wireless networks. Cognitive radio is a promising technology that allows secondary users to reuse the underutilized licensed spectrum of primary users. The major challenge for spectrum sharing is to achieve high spectrum efficiency while making non-intrusive access to the licensed bands. This requires information of availability and quality of channel resources at secondary transmitters, however, is difficult to be obtained perfectly in practice. Limited channel feedback, a few bits of channel state information sent from receiver to transmitter, provides a practical approach to detect spectrum opportunities and discover channel quality. This Springer Brief investigates spectrum sharing with limited channel feedback in various cognitive radio systems, i.e., point-to-point, broadcast scheduling and ad-hoc networks. The design aim is to optimally allocate the secondary resources so as to improve the throughput of secondary users while maintaining a certain quality of service of primary users. The analytical results of optimal resource allocation are derived via optimization theory and are verified by the numerical results. The results show that the secondary performance is significantly improved by limited feedback, and is further improved by more feedback bits, more secondary receivers and more primary side information.

This research was supported under Australian Research Council's Discovery Projects funding scheme (project number DP1094194 and DP120102030).

Sydney, Australia   Zhe Wang
Sydney, Australia   Wei Zhang

Contents

Acronyms

ACK	Acknowledgement
ARQ	Automatic repeat request
ASE	Area spectral efficiency
CDI	Channel direction information
CQI	Channel quality information
CSI	Channel state information
CDF	Cumulative distribution function
DE	Differential Evolution
ELC	Efficiency loss constraint
i.i.d.	Independent and identically distributed
KKT	Karush-Kuhn-Tucker
NACK	Negative-acknowledgement
PU	Primary user
PT	Primary transmitter
PR	Primary receiver
PER	Primary exclusive region
PPP	Poisson point process
RLC	Rate loss constraint
SNR	Signal-to-noise ratio
SINR	Signal-to-interference-plus-noise ratio
s.t.	Subject to
SU	Secondary user
ST	Secondary transmitter
SR	Secondary receiver

Chapter 1
Introduction

Abstract In this chapter, we present a general review of cognitive radio and opportunistic spectrum sharing. Then, the motivations and contributions of this brief are given.

1.1 Cognitive Radio

Wireless communication plays an increasingly important role in our daily life. Wireless devices, e.g., TV, GPS and cell phones, use invisible radio waves for data transmission through the air. Radio spectrum is a scarce resource managed by government agencies, e.g., Federal Communications Commission (FCC) and Australian Communications and Media Authority (ACMA). Radio frequency bands are assigned to various licensed carriers for interference avoidance. Given limited available spectrum, the static spectrum allocation cannot accommodate the explosive growth of high data rate services. On the other hand, recent surveys reveal that most of the licensed spectrum bands are underutilized in vast ranges of temporal and geographic domains [1].

Cognitive radio, proposed by Joseph Mitola [2], improves spectrum utilization efficiency. The licensed user, also named primary user (PU), has the priority in spectrum usage. The unlicensed user, also called secondary user (SU), opportunistically reuses the underutilized spectrum bands which are defined as spectrum holes [3]: *"a spectrum hole is a band of frequencies assigned to a primary user, but, at a particular time and specific geographic location, the band is not being utilized by that user."* Generally, spectrum holes are classified into two categories: temporal spectrum hole and spatial spectrum hole [4, 5] (see Fig. 1.1). A temporal spectrum hole occurs when there is no primary transmission in a particular frequency band of interest during a piece of time period. A spatial spectrum hole exists in the geographic region that is far from primary receivers (PRs). With the advances of multiple antennas, more dimensions of spectrum holes are created, e.g., code dimension and angle dimension [6]. In this brief, we focus on exploiting temporal and spatial spectrum holes.

© The Author(s) 2015
Z. Wang, W. Zhang, *Opportunistic Spectrum Sharing in Cognitive Radio Networks*,
SpringerBriefs in Electrical and Computer Engineering, DOI 10.1007/978-3-319-15542-5_1

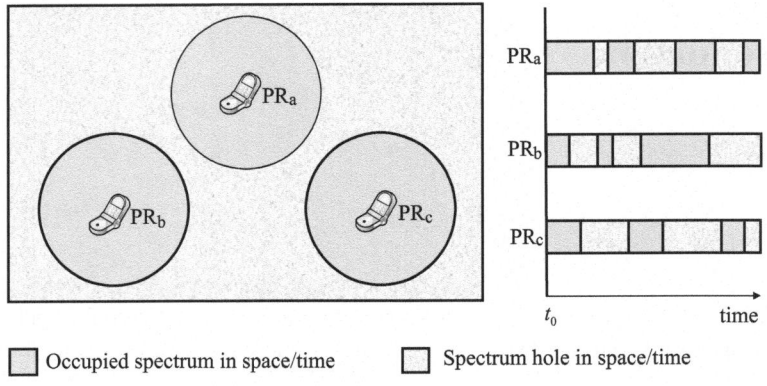

Fig. 1.1 Temporal and spatial spectrum holes

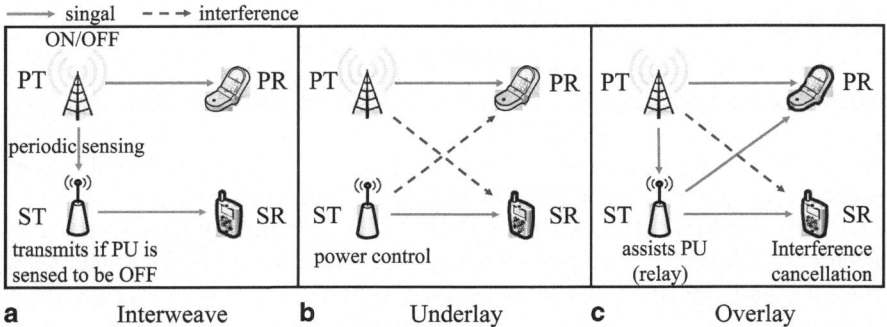

Fig. 1.2 Spectrum sharing paradigms. Primary transmitter, primary receiver, secondary transmitter and secondary receiver are denoted by PT, PR, ST, SR, respectively

1.2   Spectrum Detection and Sharing

In cognitive radio networks, the aim of SU is to discover and utilize the spectrum holes in a most efficient and non-intrusive way. Non-intrusive means PU is oblivious to SU and does not have a noticeable performance degradation.

Before accessing the spectrum, SU detects the spectrum holes via spectrum sensing [6–8], geolocation database or primary feedback signals [9–11].

Once the spectrum holes are detected, SU shares the frequency band with PU. In the literature, spectrum sharing is classified into three paradigms: interweave spectrum sharing, overlay spectrum sharing and underlay spectrum sharing, in terms of the available side information [12]. SU signals may interweave, underlay or overlay with PU signals to avoid, control or mitigate the interference from SU to PU, as shown in Fig. 1.2.

- In interweave spectrum sharing, SU utilizes the temporal spectrum holes by avoiding simultaneous transmission with PU. SU accesses the spectrum if PU is sensed to be absent [13].
- In underlay spectrum sharing, SU may have concurrent transmissions with PU. SU explores the spatial spectrum holes by controling its transmit power not to violate the tolerable interference threshold of PU [14–24].
- In overlay spectrum sharing, SU has the side information of the codebooks and messages of PU. SU mitigates the interference in the concurrent transmission by assisting the primary transmission [25–28]. In return, SU obtains additional transmission time or bandwidth for its own transmission.

With more side information of spectrum holes, SU may adopt a mixed spectrum sharing strategy that takes advantage of the strengths of the above three paradigms.

1.3 Feedback Based Spectrum Sharing

When SU transmits simultaneously with PU, it should not impose harmful violation to the instantaneous or average quality of service (QoS) of PU, e.g., throughput and reliability. Signal-to-interference-and-noise ratio (SINR) is a performance metric to evaluate throughput and reliability. SINR at the PR is related to PT-PR and ST-PR channels which are denoted by h_{pp} and h_{sp}, respectively. Based on the amount of available channel information, the interference control is discussed as follows.

- If ST knows the instantaneous h_{pp} and h_{sp}, it adapts its transmission to the instantaneous channels so that the instantaneous and average primary QoS do not suffer a noticeable degradation.
- If ST knows statistics instead of instantaneous information of h_{pp} and h_{sp}, it adopts a constant power which does not violate the average primary QoS constraints.
- If no channel information is known at the ST, it stays beyond a certain distance away from PR.

PU is better protected if more precise channel state information (CSI) is available at ST. However, it is difficult to obtain perfect CSIT in practice. Limited feedback of CSI from PR provides ST an indirect way to assess the primary activity and the tolerable degree of interference. ST opportunistically accesses the spectrum by altering its transmission behavior according to the limited feedback. In the following, we review three feedback based spectrum sharing strategies.

1.3.1 Feedback Cooperation

PR cooperation is a straightforward method to obtain the instantaneous CSI of h_{sp} at ST [29–32]. As shown in Fig. 1.3a, the PR sends cooperative feedback of the

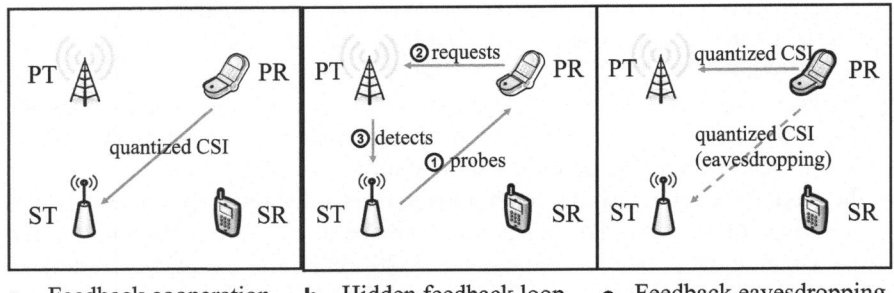

a Feedback cooperation   **b** Hidden feedback loop   **c** Feedback eavesdropping

Fig. 1.3 Limited feedback strategies

quantized channel quality information (CQI) or channel direction information (CDI) of h_{sp} to the ST. With the quantized CDI of h_{sp}, the ST computes a transmit beamformer to be orthogonal to the ST-PR channel. By transmitting within the null-space of PU, the ST reduces its interference to the PR [29]. With the quantized CQI of h_{sp}, the ST restricts its transmit power to meet the tolerable interference margin at the PR [29–32]. Though having explicit PR cooperation is a straightforward method, it may conflict with the general belief that PU is oblivious to SU.

1.3.2 Hidden Feedback Loop

In the second strategy, ST adopts the hidden feedback loop to estimate the primary SINR [33–35]. As shown in Fig. 1.3b, the ST sends a probing message to deliberately interfere the PR. If the SU signal causes a noticeable degradation to the PU performance, the PR feeds back a power-boosted request signal to the PT. Then, by detecting the power adaptation of the PT, the ST is able to know whether the PU is active or not and how much interference can be tolerated by the PR.

1.3.3 Feedback Eavesdropping

If PU is an error control system or rate adaptive system, PR sends the primary Automatic Repeat-reQuest (ARQ) signals [36–40] or quantized feedback of h_{pp} to PT [41], respectively. SU may estimate the spectrum opportunities by eavesdropping the primary feedback, as shown in Fig. 1.3c.

For the PU with ARQ, the PR sends error control signals to the PT indicating whether it has correctly received the message or not. The PT transmits a new packet or retransmits the previous packet if the PR feeds back acknowledgement (ACK) or negative-acknowledgement (NACK) in the previous time slot, respectively. For slow fading channels, the ST eavesdrops the PR control signals and takes the advantage of opportunities that arise during the PU retransmission [37]. The ST accesses the spectrum when the primary channel is in extremely good state or deep fading state [38]. For fast fading channels, the ST monitors the PU performance, e.g., outage probability, by counting the eavesdropped primary ACK/NACK during a certain length of time period and then adapts its behavior accordingly [39, 40].

In ARQ protocol, the PT makes binary decision between transmission or retransmission based on whether the CQI of h_{pp} is above or below a certain threshold. To achieve better throughput, the CQI of h_{pp} may be quantized into more than two regions and the PT may transmit with more adaptive power and rates based on the feedback [41–43]. By eavesdropping the primary feedback, the SU obtains more precise knowledge of h_{pp} and adapts its power and rate to control its interference to the PR [41].

1.4 Structure of the Brief

As discussed in the previous section, primary feedback eavesdropping is a practical and non-intrusive method for PU detection and protection. The spectral efficiency can be further enhanced if the SR also sends limited feedback of the secondary channel to the ST. This brief studies practical, non-intrusive and high spectrum efficiency spectrum sharing schemes with both primary and secondary feedback in various cognitive radio systems, i.e., point-to-point network, broadcast scheduling network and ad hoc network. The remainder of the brief is organized as follows.

In Chap. 2, we investigate the opportunistic spectrum sharing in a point-to-point cognitive network where both the PU and SU are discrete power-rate adaptive systems with limited channel feedback [44]. By receiving the secondary feedback and eavesdropping the primary feedback, the ST selects a proper power-rate pair from a pre-designed secondary quantization codebook which is designed to maximize the average throughput of the SU while not causing harmful throughput degradation to the PU. We discuss the quantization and transmission schemes of the SU in three cases with different amounts of side information of the PU interference. The results show the SU throughput is greatly improved by adapting the SU transmission to both the primary and secondary feedback. Moreover, more secondary feedback bits and more PU side information further improve the SU throughput.

In Chap. 3, we study the opportunistic spectrum sharing in a downlink cognitive network [45]. The PR and multiple SRs each sends 1-bit feedback of CQI to their corresponding transmitters. Based on the primary and secondary feedback, one of the SRs is scheduled at the ST in each fading block. We derive asymptotically optimal resource allocation that maximizes the SU throughput while protecting the PU throughput. The secondary throughput is proved to grow double logarithmically with the number of SRs.

In Chap. 4, we explore the opportunistic spectrum sharing in an ad hoc cognitive network, where both the PUs and SUs follow Poisson point processes [46]. Two spectrum sharing schemes are proposed based on the availability of primary location information. In the first scheme, the SUs do not have the location information of the PRs and access the spectrum based on the local secondary feedback. In the second scheme, the SUs have the PR location information and adapt their transmission to both the local secondary feedback and nearby primary feedback. For both schemes, we derive the closed-form optimal node density of the SU that maximizes the secondary throughput while protecting the reliability of both the PU and SU. The results show the primary location information improves the SU throughput when the PU has a high QoS requirement.

In Chap. 5, the conclusions are drawn.

References

1. Federal Communications Commission, "Spectrum policy task force report", ET Docket 02–155, Nov. 2002.
2. J. Mitola, "Cognitive radio: An integrated agent architecture for software defined radio," Ph.D. dissertation, KTH, Stockholm, Sweden, 2000.
3. P. Kolodzy et al. "Next generation communications: Kickoff meeting," in *Proc. DARPA*, Oct. 17, 2001.
4. R. Tandra, S. M. Mishra and A. Sahai, "What is a spectrum hole and what does it take to recognize one?" *Proc. IEEE*, vol. 97, pp. 824–848, May 2009.
5. J. Ma, G. Y. Li, and B. H. Juang, "Signal processing in cognitive radio," *Proc. IEEE*, vol. 97, pp. 805–823, May 2009.
6. T. Yucek and H. Arslan, "A survey of spectrum sensing algorithms for cognitive radio applications," *IEEE Commun. Survey & Tutorials*, vol. 11, pp. 116–130, First Quater 2009.
7. W. Zhang and K. B. Letaief, "Cooperative spectrum sensing with transmit and relay diversity in cognitive radio networks," *IEEE Trans. Wireless Commun.,* vol. 7, pp. 4761–4766, Dec. 2008.
8. I. F. Akyildiz, B. F. Lo, and R. Balakrishnan "Cooperative spectrum sensing in cognitive radio networks: A survey," *Physical Commun. (Elsevier) Journal,* vol. 4, pp. 40–62, Mar. 2011.
9. Federal Communications Commission, "Notice of proposed rule making: Unlicensed operation in the TV broadcast bands," ET Docket No. 04–186 (FCC 04–113), May 2004.
10. M. Marcus, "Unlicensed cognitive sharing of TV spectrum: the controversy at the federal communications commission," *IEEE Commun. Mag.*, vol. 43, no. 5, pp. 24–25, 2005.
11. Y. Zhao, L. Morales, J. Gaeddert, K. K. Bae, J.-S. Um, and J. H. Reed, "Applying radio environment maps to cognitive wireless regional area networks," in Proc. *IEEE Int. Symposium on New Frontiers in Dynamic Spectrum Access Networks*, Dublin, Ireland, Apr. 2007, pp. 115–118.
12. A. Goldsmith, S. A. Jafar, I. Maric, and S. Srinivasa, "Breaking spectrum gridlock with cognitive radios: An information theoretic perspective," *Proc. IEEE*, vol. 97, pp. 894–914, May 2009.
13. Y. C. Liang, Y. Zeng, E. Peh and A. T. Hoang, "Sensing-throughput tradeoff for cognitive radio networks", *IEEE Trans. Wireless Commun.,* vol. 7, pp. 1326–1337, Apr. 2008.
14. M. Gastpar, "On capacity under receive and spatial spectrum-sharing constraints," *IEEE Trans. Inf. Theory*, vol. 53, pp. 471–487, Feb. 2007.
15. A. Ghasemi and E. S. Sousa, "Fundamental limits of spectrum-sharing in fading environments," *IEEE Trans. Wireless Commun.*, vol. 6, pp. 649–658, Feb. 2007.
16. Q. Zhao and B. Sadler, "A survey of dynamic spectrum access," *IEEE Signal Process. Mag.,* vol. 24, pp. 79–89, May 2007.

17. X. Kang, Y. C. Liang, A. Nallanathan, H. K. Garg, and R. Zhang, "Optimal power allocation for fading channels in cognitive radio networks: Ergodic capacity and outage capacity," *IEEE Trans. Wireless Commun.*, vol. 8, pp. 940–950, Feb. 2009.
18. R. Zhang, "On peak versus average interference power constraints for protecting primary users in cognitive radio networks," *IEEE Trans. Wireless Commun.*, vol. 8, pp. 2112–2120, Apr. 2009.
19. R. Zhang, Y. C. Liang, and S. Cui, "Dynamic resource allocation in cognitive radio networks," *IEEE Signal Process. Mag.*, vol. 27, pp. 102–114, May 2010.
20. M. Vu, N. Devroye, M. Sharif, and V. Tarokh, "Scaling laws of cognitive networks," in *Proc. Int. Conf. Cognitive Radio Oriented Wireless Netw. and Commun. (Crowncom 2007)*, Orlando, Florida, USA, Jul. 31-Aug. 3, 2007.
21. L. Musavian and S. Aissa, "Capacity of spectrum-sharing channels with minimum-rate requirments," in *Proc. IEEE Int. Conf. Commun. (ICC 2008)*, Beijing, China, May 19–23, pp. 4639–4643, May 2008.
22. Y. Chen, G. Yu, Z. Zhang, H. H. Chen, and P. Qiu, "On cognitive radio networks with opportunistic power control strategies in fading channels," *IEEE. Trans. Wireless Commun.*, vol. 7, pp. 2752–2761, Jul. 2008.
23. X. Kang, R. Zhang, Y.-C. Liang, and H. K. Garg, "Optimal power allocation strategies for fading cognitive radio channels with primary user outage constraint," *IEEE J. Sel. Areas Commun.*, vol. 29, pp. 374–383, Feb. 2011.
24. R. Zhang, "Optimal power control over fading cognitive radio channel by exploiting primary user CSI," in *Proc. IEEE Global Commun. Conf. (GLOBECOM)*, New Orleans, USA, Nov. 2008.
25. S. Srinivasa and S. A. Jafar, "The throughput potential of cognitive radio: A theoretical perspective," *IEEE Commun. Mag.*, vol. 45, pp. 73–79, May 2007.
26. Y. Han, A. Pandharipande, and S. H. Ting, "Cooperative decode-andforward relaying for secondary spectrum access," *IEEE Trans. Wireless Commun.*, vol. 8, pp. 4945–4950, Oct. 2009.
27. Y. Han, S. H. Ting, and A. Pandharipande, "Cooperative spectrum sharing protocol with secondary user selection," *IEEE Trans. Wireless Commun.*, vol. 9, pp. 2914–2923, Sep. 2010.
28. C. Zhai, W. Zhang, and P. C. Ching, "Cooperative spectrum sharing based on two-path successive relaying," *IEEE Trans. Wireless Commun.*, vol. 61, pp. 2260–2270, Jun. 2013.
29. K. Huang and R. Zhang, "Cooperative feedback for multiantenna cognitive radio networks," *IEEE Trans. Signal Process.*, vol. 59, pp. 747–758, Feb. 2011.
30. Y. He and S. Dey, "Power allocation in spectrum sharing cognitive radio networks with quantized channel information," *IEEE Trans. Commun.*, vol. 59, pp. 1644–1656, Jun. 2011.
31. Y. He and S. Dey, "Throughput maximization in cognitive radio under peak interference constraints with limited feedback," *IEEE Trans. Veh. Technol.*, vol. 61, pp. 1287–1305, Mar. 2012.
32. M. M. Abdallah, A. H. Salem, M.-S. Alouini, and K. A. Qaraqe, "Adaptive discrete rate and power transmission for spectrum sharing systems," *IEEE Trans. Wireless Commun.*, vol. 11, pp. 1283–1289, Apr. 2012.
33. R. Zhang and Y. C. Liang, "Exploiting hidden power-feedback loops for cognitive radio," in *Proc. IEEE Int. Symp. New Frontiers Dynamic Spectrum Access Netw. (DySPAN 2008)*, Chicago, Ilinois, Oct. 14–17, 2008.
34. R. Zhang, "On active learning and supervised transmission of spectrum sharing based cognitive radios by exploiting hidden primary radio feedback," *IEEE Trans. Commun.*, vol. 58, pp. 2960–2970, Oct. 2010.
35. G. Zhao, Y. (G.) Li, C. Yang, and J. Ma, "Proactive detection of spectrum holes in cognitive radio," in *Proc. IEEE Int. Conf. Commun. (ICC 2009)*, Dresden, Germany, June 14–18, 2009.
36. M. Levorato, U. Mitra, and M. Zorzi, "Cognitive interference management in retransmission-based wireless networks," *IEEE Trans. Inf. Theory*, vol. 58, pp. 3023–3046, May 2012.

37. R. A. Tannious and A. Nosratinia, "Cognitive radio protocols based on exploiting hybrid ARQ retransmission," *IEEE Trans. Wireless Commun.*, vol. 9, pp. 2833–2841, Sep. 2010.
38. J. C. F. Li, W. Zhang, A. Nosratinia, and J. Yuan, "SHARP: Spectrum harvesting with ARQ retransmission and probing in cognitive radio," *IEEE Trans. Commun.*, vol. 61, pp. 951–960, Mar. 2013.
39. S. Huang, X. Liu, and Z. Ding, "Decentralized cognitive radio control based on inference from primary link control information," *IEEE J. Sel. Areas Commun.*, vol. 29, pp. 394–406, Feb. 2011.
40. K. Eswaran, M. Gastpar, and K. Ramchandran, "Cognitive radio through primary control feedback," *IEEE J. Sel. Areas Commun.*, vol. 29, pp. 384–393, Feb. 2011.
41. J. C. F. Li, W. Zhang, and J. Yuan, "Opportunistic spectrum sharing in cognitive radio networks based on primary limited feedback," *IEEE Trans. Commun.*, vol. 59, pp. 3272–3277, Dec. 2011.
42. L. Lin, R. D. Yates, and P. Spasojevic, "Adaptive transmission with discrete code rates and power levels," *IEEE Trans. Commun.*, vol. 51, pp. 2115–2125, Dec. 2003.
43. T. T. Kim and M. Skoglund, "On the expected rate of slowly fading channels with quantized side information," *IEEE Trans. Commun.*, vol. 55, pp. 820–829, Apr. 2007.
44. Z. Wang and W. Zhang, "Spectrum sharing with limited channel feedback," *IEEE Transactions on Wireless Communications*, vol. 12, pp. 2524–2532, May 2013.
45. Z. Wang and W. Zhang, "Exploiting multiuser diversity with 1-bit feedback for spectrum sharing," *IEEE Transactions on Communications*, vol. 62, pp. 29–40, Jan. 2014.
46. Z. Wang and W. Zhang, "Opportunistic spectrum sharing with limited feedback in Poisson cognitive radio networks," *IEEE Trans. Wireless Commun.*, vol. 13, no. 12, pp. 7098–7109, Dec. 2014.

Chapter 2
Cognitive Point-to-Point Network with Limited Feedback

Abstract In this chapter, we study an opportunistic spectrum sharing scheme in a point-to-point network, where both the primary user and secondary user are power-rate adaptive systems with a few bits of quantized channel feedback from the receivers. The secondary user detects the spectrum opportunities by overhearing the primary feedback and receiving the secondary feedback, and then adapts its power and rate accordingly. The secondary quantization codebook is designed offline by maximizing the secondary average throughput while not causing noticeable degradation to the primary throughput. The secondary quantization and transmission schemes are discussed in three cases when different primary interference information is available at the secondary receiver. Numerical results show that the secondary throughput is greatly improved by the introduction of even 1-bit feedback. Moreover, the secondary throughput increases with more secondary feedback bits or more primary side information.

Keywords Cognitive radio · Spectrum sharing · Quantized channel feedback · Rate loss constraint

2.1 Introduction

In cognitive radio network, secondary user (SU) aims at detecting and utilizing the spectrum holes of primary user (PU) in a non-intrusive and efficient way. Specifically,

- *PU is oblivious to SU*: PU neither is aware of the existence of SU nor cooperates with SU.
- *SU is non-intrusive to PU*: The interference from SU to PU should not cause a harmful degradation to the PU performance.
- *High spectrum efficiency*: SU achieves good performance.

With these three targets in mind, we study an opportunistic spectrum sharing scheme in the point-to-point network in this chapter [1, 2]. As shown in Fig. 2.1, both the PU

© [2013] IEEE. Reprinted, with permission, from [Z. Wang and W. Zhang, "Spectrum sharing with limited channel feedback," *IEEE Trans. Wireless Commun.*, vol. 12, pp. 2524–2532, May 2013.]

© The Author(s) 2015

Z. Wang, W. Zhang, *Opportunistic Spectrum Sharing in Cognitive Radio Networks*,
SpringerBriefs in Electrical and Computer Engineering, DOI 10.1007/978-3-319-15542-5_2

Fig. 2.1 Spectrum learning and adaptation

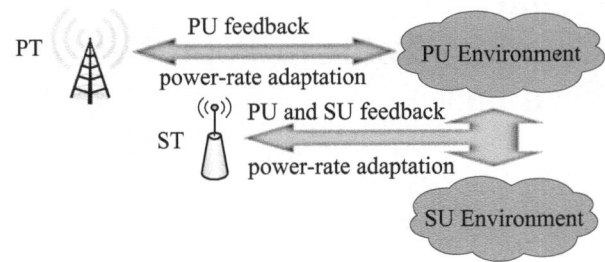

and SU are rate adaptive systems with limited channel feedback from the receivers. The primary transmitter (PT) obtains the primary channel information via primary feedback, and then adapts its power and rate accordingly. Since the PU is oblivious to the SU, the primary receiver (PR) neither receives the secondary feedback nor sends cooperative feedback to the SU. The secondary transmitter (ST) detects the spectrum holes non-intrusively, i.e., by eavesdropping the primary channel feedback, and obtains the secondary channel information via the secondary feedback. Then the ST makes efficient spectrum access by adapting its power and rate to the both the primary and secondary feedback. The secondary power and rate are designed not to cause harmful degradation to the average throughput of the PU, which guarantees the non-intrusive spectrum access.

The interference from the PT to secondary receiver (SR) was neglected in previous work [3, 4]. In this work, we consider that the SR may be able to cancel the interference from the PT based on the priori of the primary interference, i.e., primary signal and interference channel. The secondary sharing schemes are discussed in three cases when the SR has full, partial and no priori of the primary interference.

- In Case I, both the primary signal and PT-SR channel are known at the SR;
- In Case II, the primary signal is unknown but PT-SR channel is known at the SR;
- In Case III, neither the primary signal nor PT-SR channel is known at the SR.

Firstly, we propose the quantization, feedback and transmission schemes for the three cases. We establish the primary and secondary quantization codebook with power and rate that are to be obtained in the next step. Secondly, we optimize the power and rate in the primary and secondary quantization codebook offline. The average primary throughput is maximized under the average power constraint of the PU. The average secondary throughput is maximized under the average power constraint of the SU and average rate loss constraint of the PU. Thirdly, we use the optimal quantization codebook online for opportunistic spectrum sharing.

The rest of the chapter is organized as follows. In Sect. 2.2, system model for point-to-point spectrum sharing is introduced. In Sects. 2.3 and 2.4, primary and secondary quantization codebooks are designed, respectively. In Sect. 2.5, numerical results are discussed. Finally, summary is given in Sect. 2.6.

Fig. 2.2 System model of point-to-point spectrum sharing with limited feedback. Signals, interference, primary feedback and secondary feedback are represented by *solid, long dashed, dotted,* and *short dashed arrows,* respectively

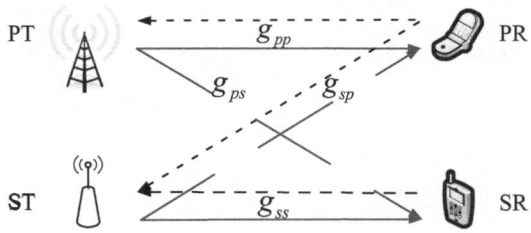

2.2   System Model

As shown in Fig. 2.2, a pair of ST and SR share the same frequency band with a pair of PT and PR. Denote g_{pp}, g_{ps}, g_{sp} and g_{ss} the instantaneous channel power gains of the PT-PR, PT-SR, ST-PR and ST-SR channels, which follow exponential distribution with mean of $\bar{g}_{pp}$, $\bar{g}_{ps}$, $\bar{g}_{sp}$ and $\bar{g}_{ss}$, respectively. Assume the PT and ST have no information about the instantaneous channels, but know the distributions and mean values of the channel power gains. The PR and SR are assumed to obtain perfect knowledge of g_{pp} and g_{ss}, respectively, through the training process and send a few bits of feedback of the quantized channel power gains to their corresponding transmitters. The additive white Gaussian noise is assumed to be with zero mean and variance N_0. Consider a block fading scenario where channel remains the same over one block but differs across blocks.

2.3   PU Quantization Codebook

The PT adapts its transmission to the quantized feedback of g_{pp} from the PR. g_{pp} is quantized it into N regions, i.e., $[0, \gamma_1^p)$, $[\gamma_1^p, \gamma_2^p)$, $\cdots$, $[\gamma_{N-1}^p, +\infty)$ (let $\gamma_N^p = +\infty$), where γ_n^p ($n = 1, 2, \cdots, N-1$) is the boundary of the primary quantization regions. Assume the PR obtains the perfect knowledge of the instantaneous g_{pp} after channel training. If $g_{pp} \in [\gamma_n^p, \gamma_{n+1}^p)$, the PR sends index n back to the PT. The total number of feedback bits for the PU is $\lceil \log_2 N \rceil$. When receiving the primary quantization index n ($n = 0, 1, \cdots, N-1$), the PT selects the corresponding power P_n^p and transmission rate R_n^p from the n-th row in the primary quantization codebook as shown in Table 2.1, and uses them in the current block.

Next, we solve the optimal power-rate allocation in Table 2.1. This problem is the same as that in the stand-alone point-to-point network [5, 6] and had also been discussed in [4]. To ensure the non-zero rate in the first quantization region $[0, \gamma_1^p)$, an outage threshold $\gamma_0^p > 0$ is introduced. For each n ($n = 0, 1, \cdots, N-1$), the discrete transmission rate is a function of transmit power and quantization threshold, i.e.,

$$R_n^p = \log\left(1 + \frac{\gamma_n^p P_n^p}{N_0}\right). \tag{2.1}$$

Table 2.1 PU quantization
codebook for the
point-to-point network

g_{pp} region	PU index	PU power and rate
$[0, \gamma_1^p)$	0	P_0^p, R_0^p
...	...	...
$[\gamma_n^p, \gamma_{n+1}^p)$	n	P_n^p, R_n^p
...	...	...
$[\gamma_{N-1}^p, +\infty)$	$N-1$	P_{N-1}^p, R_{N-1}^p

Note $\log(\cdot)$ stands for natural logarithm function throughout this chapter. Given $\lceil \log_2 N \rceil$ bits of feedback, the optimal γ_n^p and P_n^p ($n = 0, 1, \cdots, N-1$) are obtained by maximizing the average PU throughput $\bar{R}^p$ subject to the average power constraint of the PU P_{th}^p.

$$P2.1: \max_{\{\gamma_n^p, P_n^p\}} \bar{R}^p = \sum_{n=0}^{N-1} G_n^p R_n^p \tag{2.2}$$

$$\text{s.t.} \quad F^p(\gamma_1^p)P_0^p + \sum_{n=1}^{N-1} G_n^p P_n^p \leq P_{th}^p, \tag{2.3}$$

where $F^p(x) = 1 - \exp\left(-x/\bar{g}_{pp}\right)$ is the cumulative distribution function (CDF) of g_{pp}, and $G_n^p = F^p(\gamma_{n+1}^p) - F^p(\gamma_n^p) = \exp\left(-\gamma_n^p/\bar{g}_{pp}\right) - \exp\left(-\gamma_{n+1}^p/\bar{g}_{pp}\right)$ is the probability that $g_{pp} \in [\gamma_n^p, \gamma_{n+1}^p)$ for $n = 0, 1, \cdots, N-1$. Problem $P2.1$ can be solved by Algorithm 1 in [6].

2.4   SU Quantization Codebook

In this section, we discuss the quantization and feedback schemes of the SU and design the optimal power and rate that maximize the average SU throughput.

Assume that the SU has the knowledge of the primary quantization codebook and is able to overhear the primary feedback. To make non-intrusive and efficient spectrum access, the SU adapts its feedback and transmission to the primary environment, i.e., primary channel quality. For each primary index n, the secondary channel power gain is quantized into M regions, i.e., $[0, \gamma_{n,1}^s)$, $[\gamma_{n,1}^s, \gamma_{n,2}^s), \cdots, [\gamma_{n,(M-1)}^s, +\infty)$ (let $\gamma_{n,M}^s = +\infty$), where $\gamma_{n,m}^s$ ($m = 1, 2, \cdots, M-1$) indicates the secondary quantization threshold. When the secondary channel power gain falls into a certain region, i.e., $[\gamma_{n,m}^s, \gamma_{n,(m+1)}^s)$, index m is sent from the SR to the ST. The total number of feedback bits for the SU is $\lceil \log_2 M \rceil$. When the ST receives the secondary index m ($m = 0, 1, \cdots, M-1$) and overhears the primary index n, it selects the corresponding power $P_{n,m}^s$ and rate $R_{n,m}^s$ from the n-th row and m-th column in the secondary quantization codebook as shown in Table 2.2, and uses them to the end of the current block. As mentioned, both the primary and secondary quantization codebooks are designed offline.

Table 2.2 SU quantization codebook for the point-to-point network

PU index	SU index				
	0	$\cdots$	m	$\cdots$	$M-1$
0	$[0, \gamma^s_{0,1})$ $P^s_{0,0}$ $R^s_{0,0}$	$\cdots$	$[\gamma^s_{0,m}, \gamma^s_{0,(m+1)})$ $P^s_{0,m}$ $R^s_{0,m}$	$\cdots$	$[\gamma^s_{0,(M-1)}, +\infty)$ $P^s_{0,(M-1)}$ $R^s_{0,(M-1)}$
$\cdots$	$\cdots$	$\cdots$	$\cdots$	$\cdots$	$\cdots$
n	$[0, \gamma^s_{n,1})$ $P^s_{n,0}$ $R^s_{n,0}$	$\cdots$	$[\gamma^s_{n,m}, \gamma^s_{n,(m+1)})$ $P^s_{n,m}$ $R^s_{n,m}$	$\cdots$	$[\gamma^s_{n,(M-1)}, +\infty)$ $P^s_{n,(M-1)}$ $R^s_{n,(M-1)}$
$\cdots$	$\cdots$	$\cdots$	$\cdots$	$\cdots$	$\cdots$
$N-1$	$[0, \gamma^s_{(N-1),1})$ $P^s_{(N-1),0}$ $R^s_{(N-1),0}$	$\cdots$	$[\gamma^s_{(N-1),m}, \gamma^s_{(N-1),(m+1)})$ $P^s_{(N-1),m}$ $R^s_{(N-1),m}$	$\cdots$	$[\gamma^s_{(N-1),(M-1)}, +\infty)$ $P^s_{(N-1),(M-1)}$ $R^s_{(N-1),(M-1)}$

Define the relative degradation of the primary throughput due to the existence of the SU as the primary rate loss ratio, i.e.,

$$p^p_{RL} = 1 - \underline{R}^p / \bar{R}^p, \tag{2.4}$$

where $\underline{R}^p$ and $\bar{R}^p$ are the average PU throughput with and without the existence of the SU, respectively. $\bar{R}^p$ was given in (2.2) and $\underline{R}^p$ will be derived in Sects. 2.4.1, 2.4.2 and 2.4.3 for the three cases. To protect the average throughput of the PU, the primary rate loss ratio is constrained within the rate loss constraint (RLC), i.e., $p^p_{RL} \leq r^p_{RL}$. It was claimed in [7] that RLC outperforms that of the instantaneous interference constraint in terms of spectrum utilization efficiency.

Next, we design the optimal secondary quantization thresholds $\gamma^s_{n,m}$ and transmit power $P^s_{n,m}$ ($n = 0, 1, 2, \cdots, N-1$ and $m = 0, 1, 2, \cdots, M-1$) in Table 2.2 by maximizing the average secondary throughput $\bar{R}^s$ subject to the primary RLC r^p_{RL} and the average secondary power constraint P^s_{th}.

$$P2.2: \max_{\{\gamma^s_{n,m}, P^s_{n,m}\}} \bar{R}^s \tag{2.5}$$

$$\text{s.t. } p^p_{RL} \leq r^p_{RL} \tag{2.6}$$

$$\bar{P}^s \leq P^s_{th}, \tag{2.7}$$

where $\bar{P}^s$ is the average secondary transmit power.

Since the PU is oblivious to the SU, it does not control its interference to the SU. However, the SR may cancel the interference from the PT based on the availability of the primary signal and interference channel gain g_{ps}. In the following discussions, the secondary quantization codebook design are discussed in three cases in terms of the priori information of the primary interference. In Case I, both the primary signal and g_{ps} are known at the SR. In Case II, the primary signal is unknown but g_{ps} is known at the SR. In Case III, neither the primary signal nor g_{ps} is known at the SR. For each case, $\bar{R}^s$, $\bar{P}^s$ and p_{RL}^p in $P2.2$ will be derived.

2.4.1 Case I: Full Priori of PU Interference

In this case, both the primary signal and g_{ps} are known at the SR. The interference from the PU to SU can be fully canceled. The secondary quantization thresholds are based on g_{ss} only. g_{ss} is quantized into M regions, i.e., $[0, \gamma_1^s)$, $[\gamma_1^s, \gamma_2^s)$, $\cdots$, $[\gamma_{M-1}^s, \infty)$, where γ_m^s ($m = 1, 2, \cdots, M-1$) indicate the quantization boundaries. Since the secondary quantization is not related to the PU, we have $\gamma_{1,m}^s = \gamma_{2,m}^s = \cdots = \gamma_{(N-1),m}^s = \gamma_m^s$ for each secondary index m in Table 2.2. When the ST overhears the primary index n and receives secondary index m, it selects power $P_{n,m}^s$ and transmission rate $R_{n,m}^s$ from the secondary quantization codebook until the end of current block, where $R_{n,m}^s = \log\left(1 + P_{n,m}^s \gamma_m^s / N_0\right)$ for $n = 0, 1, \cdots, N-1$ and $m = 0, 1, \cdots, M-1$. An outage threshold $\gamma_0^s > 0$ is introduced to guarantee the non-zero rate in $[0, \gamma_1^s)$. In the following discussions, we derive $\bar{R}^s$, $\bar{P}^s$ and $\underline{R}^p$ in $P2.2$.

2.4.1.1 Average SU Throughput

Define the CDF of g_{ss} by $F^s(x) = 1 - \exp\left(-x/\bar{g}_{ss}\right)$. $G_m^s = F^s\left(\gamma_{m+1}^s\right) - F^s\left(\gamma_m^s\right)$ is the probability that $g_{ss} \in [\gamma_m^s, \gamma_{m+1}^s)$. The secondary transmission is successful if the secondary channel capacity $C_{n,m}^s = \log\left(1 + P_{n,m}^s g_{ss}/N_0\right)$ exceeds the target rate $R_{n,m}^s$, which implies $g_{ss} \geq \gamma_m^s$. Thus, G_m^s is the joint probability of the event that the ST transmits with $R_{n,m}^s$ and the event that the transmission is successful. As discussed, $F^p(\gamma_1^p)$ is the probability that $g_{pp} \in [0, \gamma_1^p)$ and $G_n^p = F^p(\gamma_{n+1}^p) - F^p(\gamma_n^p)$ is the probability that $g_{pp} \in [\gamma_n^p, \gamma_{n+1}^p)$. Given primary index 0 and secondary index m, the achievable secondary rate is $F^p(\gamma_1^p)G_m^s R_{0,m}^s$. Given primary index n and secondary index m, the achievable secondary rate is $G_n^p G_m^s R_{n,m}^s$. Considering all N PU index and M SU index in Table 2.2, the average SU throughput is given by

$$\bar{R}^s = F^p(\gamma_1^p)\left[\sum_{m=0}^{M-1} G_m^s R_{0,m}^s\right] + \sum_{n=1}^{N-1} G_n^p \left[\sum_{m=0}^{M-1} G_m^s R_{n,m}^s\right]. \tag{2.8}$$

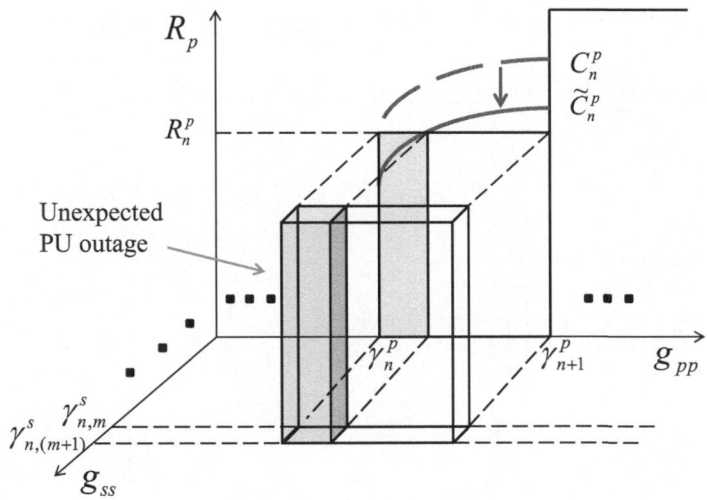

Fig. 2.3 Primary rate loss due to the existence of SU

2.4.1.2   Average SU Power

The average transmit power of the SU is given by

$$\bar{P}^s = F^p(\gamma_1^p)\left[F^s(\gamma_1^s)P_{0,0}^s + \sum_{m=1}^{M-1} G_m^s P_{0,m}^s\right] + \sum_{n=1}^{N-1} G_n^p\left[F^s(\gamma_1^s)P_{n,0}^s + \sum_{m=1}^{M-1} G_m^s P_{n,m}^s\right].$$

(2.9)

2.4.1.3   PU Rate Loss Ratio

As discussed, $\bar{R}^p$ was given in (2.2). To obtain p_{RL}^p, $\underline{R}^p$ is derived. If $g_{pp} \in [\gamma_n^p, \gamma_{n+1}^p)$, the PT transmits with the power P_n^p and transmission rate R_n^p, where $R_n^p = \log\left(1 + P_n^p \gamma_n^p / N_0\right)$.

- In the absence of the SU, the primary channel capacity is given by $C_n^p = \log\left(1 + P_n^p g_{pp}/N_0\right)$. The primary transmission is always successful ($C_n^p \geq R_n^p$) since $g_{pp} \geq \gamma_n^p$.
- In the presence of the SU, the primary channel capacity is given by $\tilde{C}_n^p = \log\left(1 + P_n^p \tilde{g}_{pp}/N_0\right)$, where $\tilde{g}_{pp} = \frac{g_{pp}}{1+g_{sp}P_{n,m}^s/N_0}$ is the effective channel gain of the primary channel. The primary transmission is successful ($\tilde{C}_n^p \geq R_n^p$) if $\tilde{g}_{pp} \geq \gamma_n^p$. Otherwise, outage occurs. An example of the primary outage event given a pair of P_n^p and $P_{n,m}^s$ is shown in the shaded cuboid of Fig. 2.3.

Therefore, at the presence of the SU, the average primary throughput is given by

$$
\underline{R}^p = \sum_{n=0}^{N-1} R_n^p \left\{ \Pr\left(\tilde{g}_{pp} \geq \gamma_n^p, \gamma_n^p \leq g_{pp} < \gamma_{n+1}^p\right) F^s\left(\gamma_1^s\right) \right.
$$

$$
\left. + \sum_{m=1}^{M-1} \Pr\left(\tilde{g}_{pp} \geq \gamma_n^p, \gamma_n^p \leq g_{pp} < \gamma_{n+1}^p\right) G_m^s \right\}, \tag{2.10}
$$

which is further derived as

$$
\underline{R}^p = \sum_{n=0}^{N-1} R_n^p \left\{ \left[e^{-\frac{\gamma_n^p}{g_{pp}}} - e^{-\frac{\gamma_{n+1}^p}{g_{pp}}} + \frac{1}{1 + \frac{\tilde{g}_{pp} N_0}{\gamma_n^p \tilde{g}_{sp} P_{n,0}^s}} \right. \right.
$$

$$
\left. \left. \times \left(e^{-\frac{\gamma_{n+1}^p}{g_{pp}} + \frac{N_0}{\tilde{g}_{sp} P_{n,0}^s}\left(1 - \frac{\gamma_{n+1}^p}{\gamma_n^p}\right)} - e^{-\frac{\gamma_n^p}{g_{pp}}} \right) \right] F^s\left(\gamma_1^s\right) \right.
$$

$$
+ \sum_{m=1}^{M-1} \left[e^{-\frac{\gamma_n^p}{g_{pp}}} - e^{-\frac{\gamma_{n+1}^p}{g_{pp}}} + \frac{1}{1 + \frac{\tilde{g}_{pp} N_0}{\gamma_n^p \tilde{g}_{sp} P_{n,m}^s}} \right.
$$

$$
\left. \left. \times \left(e^{-\frac{\gamma_{n+1}^p}{g_{pp}} + \frac{N_0}{\tilde{g}_{sp} P_{n,m}^s}\left(1 - \frac{\gamma_{n+1}^p}{\gamma_n^p}\right)} - e^{-\frac{\gamma_n^p}{g_{pp}}} \right) \right] G_m^s \right\}. \tag{2.11}
$$

$\underline{R}^p$ is related to the average channel statistics instead of the instantaneous channel power gains. Substituting (2.11) into (2.4), we obtain the primary rate loss ratio.

2.4.2 Case II: Partial Priori of PU Interference

In this case, g_{ps} is available at the SR while the primary signal is not. The primary interference is treated as noise. Given the primary index n, the SR can deduce the primary transmit power P_n^p from the primary quantization codebook and then calculate the effective channel power gain $\tilde{g}_{ss} = \frac{g_{ss}}{1 + g_{ps} P_n^p / N_0}$ for the secondary channel. Different from Case I where the quantization of g_{ss} is regardless of the primary feedback, in this case, the quantization of the $\tilde{g}_{ss}$ is related to n. For each index n ($n = 0, 1, \cdots, N - 1$), $\tilde{g}_{ss}$ is quantized into M regions, i.e., $[0, \gamma_{n,1}^s), [\gamma_{n,1}^s, \gamma_{n,2}^s), \cdots, [\gamma_{n,(M-1)}^s, +\infty)$, where $\gamma_{n,m}^s$ ($m = 1, 2, \cdots, M - 1$) is the secondary quantization threshold. If $\tilde{g}_{ss} \in [\gamma_{n,m}^s, \gamma_{n,(m+1)}^s)$, the SR feeds back index m to the ST. Then, the ST selects power $P_{n,m}^s$ and rate $R_{n,m}^s = \log\left(1 + P_{n,m}^s \gamma_{n,m}^s / N_0\right)$ from Table 2.2 for $n = 0, 1, \cdots, N - 1$ and $m = 0, 1, \cdots, M - 1$. The outage thresholds $\gamma_{n,0}^s > 0$ is introduced to guarantee the non-zero rate in $[0, \gamma_{n,1}^s)$. The CDF of $\tilde{g}_{ss}$ is given by

$$\tilde{F}^s(x) = 1 - \frac{\bar{g}_{ss} N_0}{\bar{g}_{ss} N_0 + P_n^p \bar{g}_{ps} x} \exp\left(-\frac{x}{\bar{g}_{ss}}\right). \tag{2.12}$$

The probabilities of $\tilde{g}_{ss} \in [\gamma_{n,m}^s, \gamma_{n,(m+1)}^s)$ is represented as $\tilde{G}_{n,m}^s = \tilde{F}^s\left(\gamma_{n,(m+1)}^s\right) - \tilde{F}^s\left(\gamma_{n,m}^s\right)$.

2.4.2.1 Average SU Throughput

The average secondary throughput is given by

$$\bar{R}^s = F^p(\gamma_1^p) \sum_{m=0}^{M-1} \tilde{G}_{0,m}^s R_{0,m}^s + \sum_{n=1}^{N-1} G_n^p \left[\sum_{m=0}^{M-1} \tilde{G}_{n,m}^s R_{n,m}^s\right]. \tag{2.13}$$

2.4.2.2 Average SU Power

The average transmit power of SU is given by

$$\bar{P}^s = F^p(\gamma_1^p) \left[\tilde{F}^s(\gamma_{0,1}^s) P_{0,0}^s + \sum_{m=1}^{M-1} \tilde{G}_{0,m}^s P_{0,m}^s\right]$$

$$+ \sum_{n=1}^{N-1} G_n^p \left[\tilde{F}^s(\gamma_{n,1}^s) P_{n,0}^s + \sum_{m=1}^{M-1} \tilde{G}_{n,m}^s P_{n,m}^s\right]. \tag{2.14}$$

2.4.2.3 PU Rate Loss Ratio

Similar to the discussions in Case I, the average PU rate at the presence of SU is given by

$$\underline{R}^p = \sum_{n=0}^{N-1} R_n^p \left\{\Pr\left(\tilde{g}_{pp} \geq \gamma_n^p, \gamma_n^p \leq g_{pp} < \gamma_{n+1}^p\right) \tilde{F}^s\left(\gamma_{n,1}^s\right)\right.$$

$$\left. + \sum_{m=1}^{M-1} \Pr\left(\tilde{g}_{pp} \geq \gamma_n^p, \gamma_n^p \leq g_{pp} < \gamma_{n+1}^p\right) \tilde{G}_{n,m}^s\right\}. \tag{2.15}$$

Substituting (2.15) into (2.4), we obtain the primary rate loss ratio.

2.4.3 Case III: No Priori of PU Interference

In this case, neither the primary signal nor g_{ps} is known at the SR. The primary signal is also treated as noise as in Case II. Due to the lack of g_{ps}, the SU is not able to quantize the secondary channel based on $\tilde{g}_{ss}$. Conservatively, we adopt the same method as in Case I by quantizing g_{ss} into M regions, i.e., $[0, \gamma_1^s)$, $[\gamma_1^s, \gamma_2^s)$, $\cdots$, $[\gamma_{M-1}^s, +\infty)$. Similar to Case I, we also have $\gamma_{1,m}^s = \gamma_{2,m}^s = \cdots = \gamma_{(N-1),m}^s = \gamma_m^s$ for each secondary index m in Table 2.2.

The secondary channel capacity and transmission rate are denoted by $C_{n,m}^s = \log\left(1 + P_{n,m}^s \tilde{g}_{ss}/N_0\right)$ and $R_{n,m}^s = \log\left(1 + P_{n,m}^s \gamma_m^s/N_0\right)$ for $n = 0, 1, \cdots, N-1$ and $m = 0, 1, \cdots, M-1$. Given a secondary index m, the transmission is successful if $C_{n,m}^s \geq R_{n,m}^s$. Then, the joint probability that the ST transmits with $R_{n,m}^s$ and the transmission is successful is $\Pr\left(\tilde{g}_{ss} \geq \gamma_m^s, \gamma_m^s \leq g_{ss} < \gamma_{m+1}^s\right)$. Therefore, the average secondary rate is given by

$$
\bar{R}^s = F^p\left(\gamma_1^p\right) \sum_{m=0}^{M-1} \left[\Pr\left(\tilde{g}_{ss} \geq \gamma_m^s, \gamma_m^s \leq g_{ss} < \gamma_{m+1}^s\right) R_{0,m}^s\right]
$$

$$
+ \sum_{n=1}^{N-1} \left\{ G_n^p \sum_{m=0}^{M-1} \left[\Pr\left(\tilde{g}_{ss} \geq \gamma_m^s, \gamma_m^s \leq g_{ss} < \gamma_{m+1}^s\right) R_{n,m}^s\right] \right\}. \qquad (2.16)
$$

Since the quantization region is based on g_{ss}, then $\bar{P}_s$ and $\underline{R}^p$ are regardless of the primary interference and are given by (2.9) and (2.11), respectively, as in Case I.

2.5 Numerical Results

Substituting $\bar{R}_s$, $\bar{P}_s$ and $\underline{R}^p$ for the three cases into P2.2, it is seen that P2.2 is a non-linear, non-convex and multi-modal problem which cannot be solved by most of the derivative-based optimization methods. To find the global optimal solution, Differential Evolution [8, 9], a multiple starting point, derivative-free global optimization method, is adopted with the constrained handling technique [10]. Once the global optimal secondary thresholds $\gamma_{n,m}^s$ and power $P_{n,m}^s$ in secondary quantization codebook are determined, they can be used in the online transmissions.

In this section, the numerical results for the average throughput of PU and the SU are presented. We set $\bar{g}_{pp} = 1$, $\bar{g}_{sp} = 0.5$, $\bar{g}_{ps} = 0.5$, $\bar{g}_{ss} = 4$ and $N_0 = 1$ throughout this section.

In Fig. 2.4, at the absence of the SU, the average primary throughput $\bar{R}^p$ increases as the increase of the number of primary feedback bits. The primary throughput with a few bits of feedback is almost as good as that of having perfect CQI at the PT. Moreover, $\bar{R}^p$ improves for higher primary power constraint P_{th}^p.

In the following discussions, we fix the primary power constraint $P_{th}^p = 10$ dB and the primary RLC $r_{RL}^p = 0.1$ which represents that the maximum throughput loss

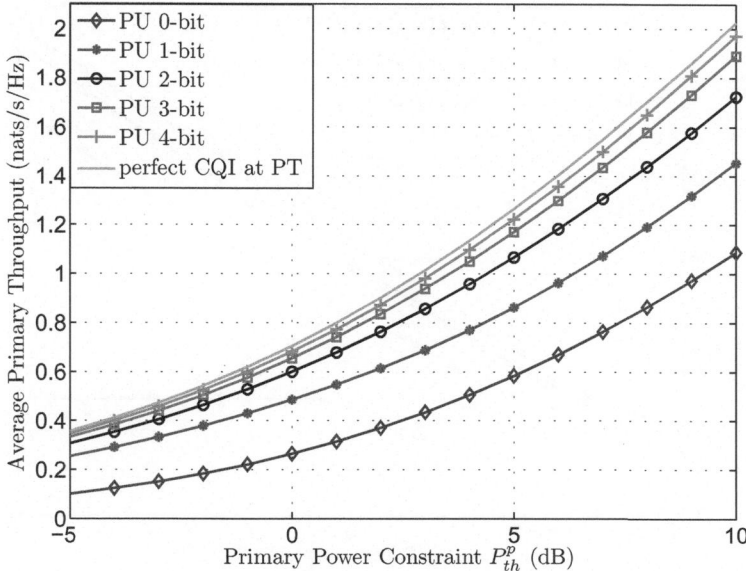

Fig. 2.4 Average throughput of the PU at the absence of the SU

tolerated by the PU is 10%. Substituting the optimal γ_n^p and P_n^p obtained in $P2.1$ into $P2.2$, we obtain $\{\gamma_{n,m}^s, P_{n,m}^s\}$ and the corresponding maximum $\bar{R}^s$ for each given P_{th}^s.

Figure 2.5 compares the average secondary throughput for Case I without and with the primary RLC, which are represented by the dashed and solid curves, respectively.

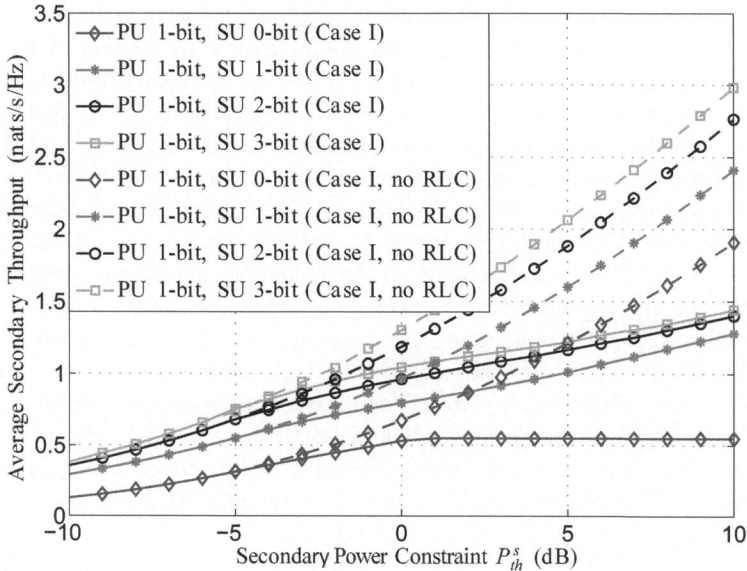

Fig. 2.5 Average throughput of the SU for Case I, where both the primary signal and g_{ps} are available at the SR

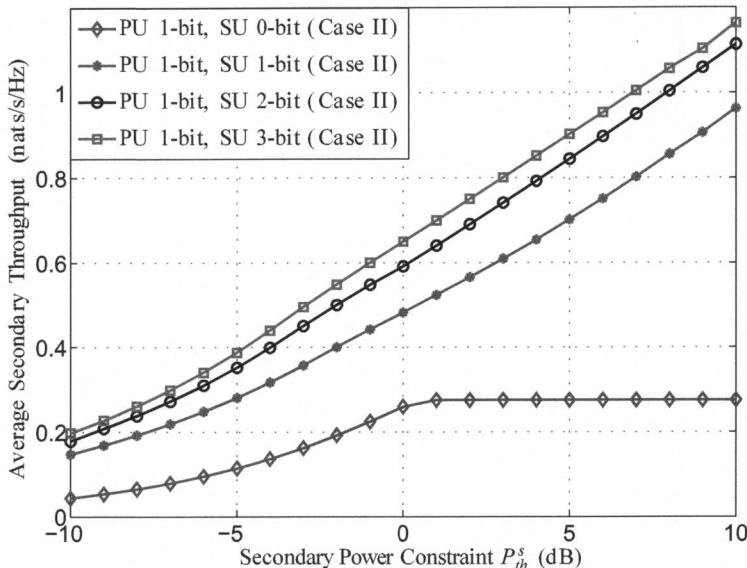

Fig. 2.6 Average throughput of the SU for Case II, where the primary signal is not available but g_{ps} is available at the SR

Without RLC, the SU is an independent system and the secondary throughput increases as P_{th}^s increases, which follows the similar trend as that of the PU in Fig. 2.4. With RLC, the increasing speed for $\bar{R}^s$ slows down as P_{th}^s increases.

Similarly, in Fig. 2.6, the secondary throughput for Case II increases rapidly with the increase of the average secondary power, but slows down when the primary RLC is activated. For both Case I and Case II, the increasing the number of secondary feedback bits results in an improvement of the average secondary throughput.

In Fig. 2.7, the secondary throughput for Case III increases with P_{th}^s first before it is flattened by the primary RLC, which is similar to the previous two cases. Though the secondary throughput improves as the introduction of 1-bit feedback, it stops increasing when more secondary bits are used. Due to the lack of information of g_{ps}, the secondary feedback is based on g_{ss}, whereas, the successful transmission is related to $\tilde{g}_{ss}$. The more partition of the secondary quantization regions may not promise a higher probability of $\Pr\left(\tilde{g}_{ss} \geq \gamma_m^s, \gamma_m^s \leq g_{ss} < \gamma_{m+1}^s\right)$. Therefore, the secondary throughput may no longer increase with more quantization bits.

Figure 2.8 compares the three cases when both the PU and SU are with 1-bit feedback. Case I outperforms Case II due to the primary interference cancelation; Case II achieves better performance than Case III thanks to the cross channel information. Generally, more side information leads to higher secondary throughput. Though Case III performs worst, there is still room for the SU to survive.

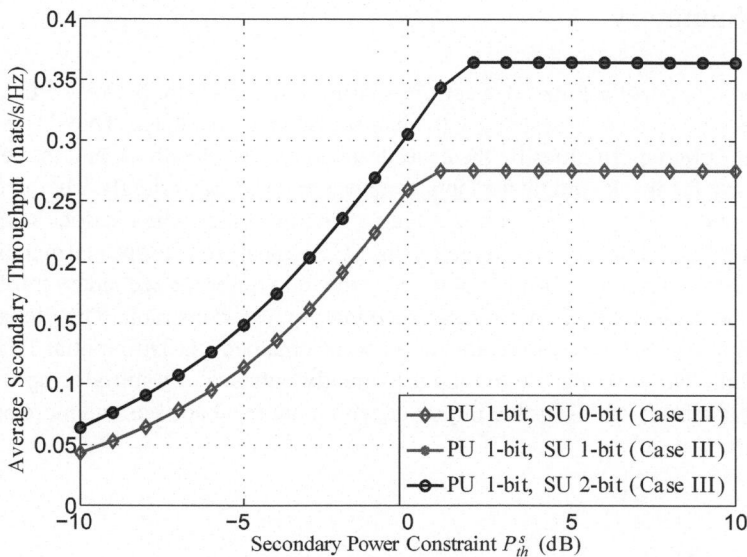

Fig. 2.7 Average throughput of the SU for Case III, where neither the primary signal nor g_{ps} is available at the SR

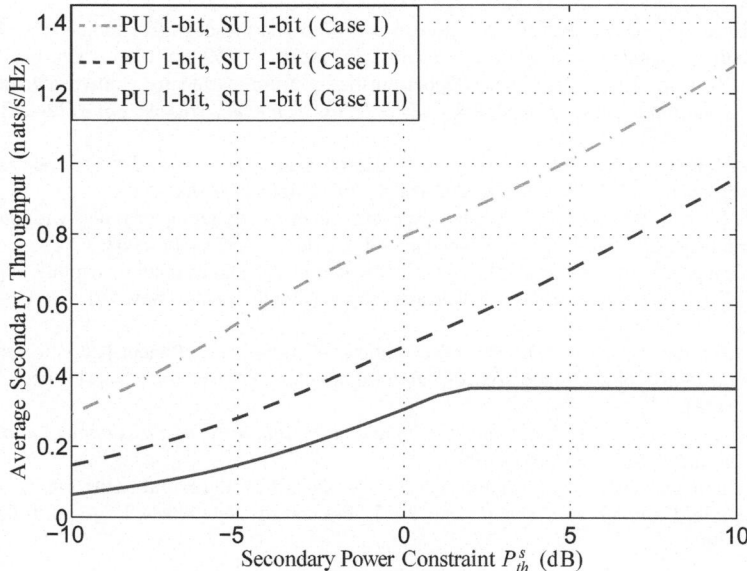

Fig. 2.8 Comparison among the three cases. For Case I, both the primary signal and g_{ps} are available at the SR. For Case II, the primary signal is not available but g_{ps} is available at the SR. For Case III, neither the primary signal nor g_{ps} is available at the SR

2.6 Summary

We studied a feedback based spectrum sharing scheme where both the PU and SU are rate-power adaptive systems with quantized channel feedback. The SU discovers the spectrum opportunities by the eavesdropped primary feedback and the received secondary feedback, and then adapts its power and rate accordingly. The secondary quantization codebook design was discussed in three cases when various side information of PU interference is available at the SU. We analyzed the optimal quantization thresholds and power allocation of the SUs that maximize the secondary throughput under the primary RLC and average secondary power constraint. The global optimal solutions of the resource allocation were obtained via Differential Evolution algorithm. The secondary throughput is greatly enhanced by introducing the secondary feedback, and is further improved with more feedback bits and more primary side-information.

References

1. Z. Wang and W. Zhang, "Spectrum sharing with primary and secondary limited feedback in cognitive radio networks," in *Proc. IEEE Global Communications Conference* (GLOBECOM 2012), Anaheim, CA, USA, Dec. 3–7, 2012, pp. 1218–1222.
2. Z. Wang and W. Zhang, "Spectrum sharing with limited channel feedback," *IEEE Trans. Wireless Commun.*, vol. 12, pp. 2524–2532, May 2013.
3. Y. He and S. Dey, "Power allocation in spectrum sharing cognitive radio networks with quantized channel information," *IEEE Trans. Commun.*, vol. 59, pp. 1644–1656, Jun. 2011.
4. J. C. F. Li, W. Zhang, and J. Yuan, "Opportunistic spectrum sharing in cognitive radio networks based on primary limited feedback," *IEEE Trans. Commun.*, vol. 59, pp. 3272–3277, Dec. 2011.
5. L. Lin, R. D. Yates, and P. Spasojevic, "Adaptive transmission with discrete code rates and power levels," *IEEE Trans. Commun.*, vol. 51, pp. 2115–2125, Dec. 2003.
6. T. T. Kim and M. Skoglund, "On the expected rate of slowly fading channels with quantized side information," *IEEE Trans. Commun.*, vol. 55, pp. 820–829, Apr. 2007.
7. R. Zhang, "Optimal power control over fading cognitive radio channel by exploiting primary user CSI," in *Proc. IEEE Global Commun. Conf. (GLOBECOM)*, New Orleans, USA, Nov. 2008.
8. R. Storn and K. Price, "Differential evolution: A simple and efficient heuristic for global optimization over continuous spaces," *Journal of Global Optimization*, vol. 11, pp. 341–359, Nov. 1997.
9. K. Price, R. Storn, and J. Lampinen, *Differential Evolution: A Practical Approach to Global Optimization*, Springer-Verlag, 2005.
10. J. Lampinen, "A constraint handling approach for the differential evolution algorithm," in *Proc. Congr. on Evolutionary Comput. (CEC 2002)*, Piscataway, New Jersey, USA, pp. 1468–1473, May 2002.

Chapter 3
Cognitive Scheduling Network with Limited Feedback

Abstract In this chapter, we consider a broadcast scheduling secondary network sharing the spectrum with a point-to-point primary network, where each receiver sends 1-bit channel feedback to its corresponding transmitter. According to the eavesdropped primary feedback and the received secondary feedback, the secondary transmitter selects a secondary receiver with one of the best instantaneous channels in each fading block and transmits to it with the adaptive rate and power. We derive the asymptotically optimal resource allocation that maximizes the secondary throughput subject to the average secondary power constraint and primary rate loss constraint. It is proved that the maximized secondary throughput grows double logarithmically with the number of receivers, which follows the same scaling law as that with the full channel quality information.

Keywords Cognitive radio · Spectrum sharing · Broadcast scheduling · Throughput scaling law · 1-bit feedback

3.1 Introduction

In the previous chapter, we have studied opportunistic spectrum sharing for the point-to-point cognitive network. As it was shown, more feedback bits improve the secondary throughput by increasing the accuracy of channel estimation. However, due to the channel fluctuation, the throughput of the secondary user (SU) may be low if its channel is in deep fading, even with perfect feedback. By introducing multiple receivers, the opportunity to find a good channel increases. This is so-called multiuser diversity [1]. To further improve the secondary throughput and reliability in Chap. 2, in this chapter, we study the opportunistic spectrum sharing in a downlink scenario and show how secondary throughput increases as the increase of the number of secondary receivers (SRs) [2].

The rest of the chapter is organized as follows. System model and transmission schemes are introduced in Sect. 3.2. In Sect. 3.3, the spectrum sharing optimization

© The Author(s) 2015

23

Z. Wang, W. Zhang, *Opportunistic Spectrum Sharing in Cognitive Radio Networks*,
SpringerBriefs in Electrical and Computer Engineering, DOI 10.1007/978-3-319-15542-5_3

Fig. 3.1 Spectrum sharing
between a secondary
broadcast scheduling network
and a pair of primary users.
Signals, interference, primary
feedback and secondary
feedback are represented by
the *solid, long dashed, dotted*
and *short dashed arrows*,
respectively

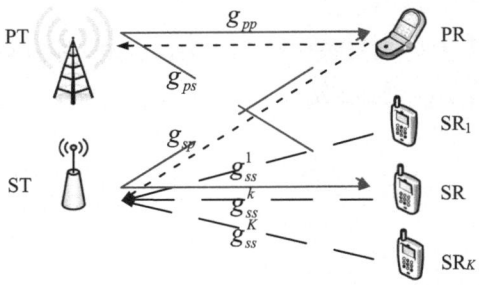

problem is given. In Sect. 3.4, the asymptotically optimal resource allocation is derived and secondary throughput scaling law is proved. In Sect. 3.5, we discuss the numerical results. Finally, the summary is drawn in Sect. 3.6.

3.2 System Model

A downlink cognitive system is shown in Fig. 3.1, where a secondary transmitter (ST) and K secondary receivers SR_k ($k = 1, 2, \cdots, K$) share the same spectrum band with a pair of primary transmitter (PT) and primary receiver (PR). Single antenna is used at each user. Denote g_{pp}, g_{ps}^k, g_{sp} and g_{ss}^k the instantaneous channel power gains of the PT-PR, PT-SR$_k$, ST-PR, ST-SR$_k$ channels, which are exponentially distributed with the mean of $\bar{g}_{pp}$, $\bar{g}_{ps}$, $\bar{g}_{sp}$ and $\bar{g}_{ss}$, respectively. Assume g_{ps}^k for $k = 1, 2, \cdots, K$ are independent and identically distributed (i.i.d.) and g_{ss}^k for $k = 1, 2, \cdots, K$ are also i.i.d. PT and ST have no information about instantaneous channels but know the distribution and mean values of the channel power gains. The perfect knowledge of g_{pp} and g_{ss}^k is assumed to be available at the PR and each SR$_k$, respectively. The PR and SRs send 1-bit feedback of CQI to their corresponding transmitters. All feedback is assumed to be perfect. Block fading is considered, where the channel power gain is assumed to be constant over each fading block. The additive white Gaussian noise at each receiver is assumed to be with zero mean and variance N_0.

3.2.1 PU Transmission Scheme

Since the primary user (PU) has higher priority to use the spectrum, it is not aware of the existence the SU. In each fading block, the PT adapts its power and rate to the primary feedback. Here we adopt similar method as that of Sect. 2.3 in Chap. 2. g_{pp} is quantized into two regions, i.e., $[0, \gamma^p)$ and $[\gamma^p, \infty)$, where γ^p is the quantization threshold. If $g_{pp} \in [0, \gamma^p)$, the PR sends index "0" back to the PT. Otherwise, it feeds back index "1". Based on the 1-bit feedback, the PT chooses the corresponding power and rate from Table 3.1. The PT keeps silent if it receives "0" and transmits with

Table 3.1 PU quantization and feedback with 1-bit feedback

PU quantization	PU feedback	PU power & rate
$g_{pp} \in [0, \gamma^p)$	"0"	$0, 0$
$g_{pp} \in [\gamma^p, \infty)$	"1"	P^p, R^p

Table 3.2 SU quantization and feedback for broadcast scheduling network with 1-bit feedback

PU feedback	SU quantization	SU feedback
"0"	$g_{ss}^k \in [0, \gamma_0^s)$	"0"
"0"	$g_{ss}^k \in [\gamma_0^s, \infty)$	"1"
"1"	$\tilde{g}_{ss}^k \in [0, \gamma_1^s)$	"0"
"1"	$\tilde{g}_{ss}^k \in [\gamma_1^s, \infty)$	"1"

power P^p and target rate $R^p = \log(1 + \gamma^p P^p / N_0)$ if it receives "1". Note $\log(\cdot)$ denotes natural logarithm function throughout this chapter. Denote $F^p(x) = 1 - \exp(-x/\bar{g}_{pp})$ the cumulative distribution function (CDF) of g_{pp}. The probabilities that $g_{pp} \in [0, \gamma^p)$ and $g_{pp} \in [\gamma^p, \infty)$ are given by $F^p = F^p(\gamma^p)$ and $G^p = 1 - F^p$, respectively. The primary quantization threshold γ^p and transmit power P^p are designed by maximizing the average primary throughput $\bar{R}^p$ subject to the average primary power constraint P_{th}^p as shown in $P3.1$.

$$P3.1 : \max_{\{\gamma^p, P^p\}} \bar{R}^p = G^p R^p \tag{3.1}$$

$$\text{s.t.} \, G^p P^p \leq P_{th}^p. \tag{3.2}$$

$P3.1$ can be solved by Algorithm 1 in [3].

3.2.2 SU Transmission Scheme

Assume the SU has the knowledge of Table 3.1 and is able to overhear the primary feedback. As the introduction of the SU, the average rate loss ratio of the PU is given by

$$p_{RL}^p = 1 - \underline{R}^p / \bar{R}^p, \tag{3.3}$$

where $\underline{R}^p$ and $\bar{R}^p$ are the average primary throughput at the presence and absence of the SU, respectively. To protect the PU, p_{RL}^p should be lower than the primary rate loss constraint (RLC) [4], i.e., $p_{RL}^p \leq r_{RL}^p$.

To make non-intrusive and efficient spectrum access, the SU adapts its feedback and transmission to the primary feedback as shown in Table 3.2. Similar to Case II in Chap. 2, g_{ps}^k is assumed to be known at the SR_k, and the primary interference is treated as noise. Denote the effective channel power gains of the $ST\text{-}SR_k$ channels by g_{ss}^k when the PU is silent and $\tilde{g}_{ss}^k = \frac{g_{ss}^k}{1 + P^p g_{ps}^k / N_0}$ when the PU is in operation. Given the overheard primary feedback, each SR_k infers if the PU is silent or operating, and then calculates g_{ss}^k or $\tilde{g}_{ss}^k$.

- If the primary feedback is "0", SR_k quantizes g_{ss}^k into two regions, i.e., $[0, \gamma_0^s)$ and $[\gamma_0^s, +\infty)$. SR_k sends secondary feedback "0" or "1" back to the ST if $g_{ss}^k \in [0, \gamma_0^s)$ or $g_{ss}^k \in [\gamma_0^s, +\infty)$, respectively.
- If the primary feedback is "1", SR_k quantizes $\tilde{g}_{ss}^k$ into regions of $[0, \gamma_1^s)$ and $[\gamma_1^s, +\infty)$. SR_k feeds back "0" or "1" to the ST if $\tilde{g}_{ss}^k \in [0, \gamma_1^s)$ or $\tilde{g}_{ss}^k \in [\gamma_1^s, +\infty)$, respectively.

If all SRs send feedback "0", the ST keeps silent. If J out of K SRs feed back "1", the ST randomly selects one SR (i.e. SR_{k^*}) from these J SRs for transmission by adapting its power and rate to both the primary and secondary feedback.

- If the primary feedback is "0", the ST transmits with power P_0^s and rate $R_0^s = \log\left(1 + \frac{P_0^s \gamma_0^s}{N_0}\right)$.
- If the primary feedback is "1", the ST transmits with power P_1^s and rate $R_1^s = \log\left(1 + \frac{P_1^s \gamma_1^s}{N_0}\right)$.

In the following two sections, we will show how to obtain the optimal γ_0^s, γ_1^s, P_0^s and P_1^s that maximize the secondary throughput offline. With the optimal solutions, the SU performs the above feedback, scheduling and transmission processes online in each fading block.

3.3 Spectrum Sharing Optimization

In $P3.2$, the problem formulation is given to determine the optimal thresholds and power of the SU in Table 3.2. The average secondary throughput $\bar{R}^s$ is maximized subject to the average secondary power constraint P_{th}^s and average primary RLC r_{RL}^p.

$$P3.2: \max_{\{\gamma_0^s, \gamma_1^s, P_0^s, P_1^s\}} \bar{R}^s \tag{3.4}$$

$$\text{s.t. } \bar{P}^s \le P_{th}^s \tag{3.5}$$

$$p_{RL}^p \le r_{RL}^p. \tag{3.6}$$

The expressions of $\bar{P}^s$, $\bar{R}^s$ and p_{RL}^p are given in the following discussions.

3.3.1 SU Power and Throughput

Let $\alpha = \frac{P^p \bar{g}_{ps}}{\bar{g}_{ss} N_0}$ and $\beta = \bar{g}_{ss}$. At the absence and presence of the PU, the CDFs of g_{ss}^k and $\tilde{g}_{ss}^k$ are $1 - \exp(-x/\beta)$ and $1 - \frac{1}{1+\alpha z} \exp(-z/\beta)$, respectively. As discussed, the ST transmits if at least one SR feeds back "1". Given primary feedback "0", the probability that the ST transmits with power P_0^s is the probability that at least one of

$g_{ss}^k \in [\gamma_0^s, \infty)$ (for $k = 1, 2, \cdots, K$), i.e.,

$$G_0^s = 1 - \left[1 - \exp\left(-\frac{\gamma_0^s}{\beta}\right)\right]^K. \tag{3.7}$$

Given primary feedback "1", the probability that the ST transmits with power P_1^s is given by

$$G_1^s = 1 - \left[1 - \frac{1}{1 + \alpha\gamma_1^s}\exp\left(-\frac{\gamma_1^s}{\beta}\right)\right]^K. \tag{3.8}$$

Since PR feeds back "0" and "1" with probabilities of F^p and G^p, respectively, the ST transmits with P_0^s and P_1^s with the probabilities of $F^p G_0^s$ and $G^p G_1^s$, respectively. The average secondary power is given by

$$\bar{P}^s = F^p G_0^s P_0^s + G^p G_1^s P_1^s. \tag{3.9}$$

Similarly, the average secondary throughput is

$$\bar{R}^s = F^p G_0^s R_0^s + G^p G_1^s R_1^s. \tag{3.10}$$

3.3.2 PU Rate Loss Ratio

Next, we discuss the relative degradation of the average primary throughput as the introduction of the SU. If $g_{pp} \in [\gamma^p, \infty)$, the PT transmits with power P^p and rate $R^p = \log\left(1 + \frac{P^p \gamma^p}{N_0}\right)$.

- In the absence of the SU, the primary transmission always succeeds in that the primary channel capacity $C^p = \log\left(1 + \frac{P^p g_{pp}}{N_0}\right) \geq R^p$.

- In the presence of the SU, the primary channel capacity is $\tilde{C}^p = \log\left(1 + \frac{P^p \tilde{g}_{pp}}{N_0}\right)$, where $\tilde{g}_{pp} = g_{pp}/\left(1 + \frac{g_{sp}P_1^s}{N_0}\right)$ is the effective channel power gain of the primary channel. The primary transmission succeeds if $\tilde{C}^p \geq R^p$, which requires $\tilde{g}_{pp} \geq \gamma^p$. Otherwise, there is an outage. In other words, the primary transmission is successful if both conditions are satisfied: $g_{pp} \in [\gamma^p, \infty)$ and $\tilde{g}_{pp} \in [\gamma^p, \infty)$.

Given the primary feedback "1", the ST transmits and not transmits with probabilities of G_1^s and $F_1^s = 1 - G_1^s$, respectively. From the above discussions, the average primary throughput is given by

$$\underline{R}^p = [F_1^s \Pr\left(g_{pp} \geq \gamma^p\right) + G_1^s \Pr\left(g_{pp} \geq \gamma^p, \tilde{g}_{pp} \geq \gamma^p\right)]R^p$$

$$= \left(1 - \frac{1}{1 + \frac{\tilde{g}_{pp}N_0}{\gamma^p \tilde{g}_{sp}P_1^s}}G_1^s\right)G^p R^p. \tag{3.11}$$

Based on (3.1), (3.3) and (3.11), the average primary rate loss ratio p_{RL}^p is

$$p_{RL}^p = \frac{G_1^s}{1 + \frac{\tilde{g}_{pp} N_0}{\gamma^p \tilde{g}_{sp} P_1^s}}. \tag{3.12}$$

According to (3.10), (3.9) and (3.12), $P3.2$ can be rewritten as $P3.3$.

$$P3.3 : \max_{\{\gamma_0^s, \gamma_1^s, P_0^s, P_1^s\}} F^p G_0^s \log\left(1 + \frac{P_0^s \gamma_0^s}{N_0}\right) + G^p G_1^s \log\left(1 + \frac{P_1^s \gamma_1^s}{N_0}\right) \tag{3.13}$$

$$\text{s.t. } F^p G_0^s P_0^s + G^p G_1^s P_1^s \leq P_{th}^s \tag{3.14}$$

$$\frac{G_1^s}{1 + \frac{\tilde{g}_{pp} N_0}{\gamma^p \tilde{g}_{sp} P_1^s}} \leq r_{RL}^p. \tag{3.15}$$

3.4 Optimal Resource Allocation

In this section, the secondary thresholds and power allocation that maximize the average throughput of the SU in $P3.3$ are derived. Moreover, the throughput scaling law of the SU is discussed.

3.4.1 Optimal Thresholds and Power

$\bar{R}^s$ is convex in P_0^s and P_1^s and unimodal in γ_0^s and γ_1^s. Given a pair of $\{P_{th}^s, r_{RL}^p\}$, there exits a unique solution of $\{\gamma_0^s, \gamma_1^s, P_0^s, P_1^s\}$. We use Karush-Kuhn-Tucker (KKT) conditions [5] to solve $P3.3$. The local optimal solution obtained via the KKT conditions is also the global optimum. However, the closed-form solution is not available. Instead, the asymptotically optimal solutions for $K \rightarrow \infty$ are derived as shown in *Theorem 3.1*. The definitions of asymptotic notations of $\Theta(\cdot)$ and $O(\cdot)$ are given as follows.

Definition 3.1 [6, Page 44] Let $\rho(K)$ and $\sigma(K)$ be asymptotically non-negative functions. $\rho(K) = \Theta(\sigma(K))$ if there exist positive constants c_1, c_2 and η such that for every $K \geq \eta, 0 \leq c_1 \sigma(K) \leq \rho(K) \leq c_2 \sigma(K)$.

Definition 3.2 [6, Page 47] Let $\rho(K)$ and $\sigma(K)$ be asymptotically non-negative functions. $\rho(K) = O(\sigma(K))$ if there exist positive constants c and η such that for every $K \geq \eta, 0 \leq \rho(K) \leq c\sigma(K)$.

Theorem 3.1 *In $P3.3$, as $K \rightarrow \infty$, the asymptotically optimal thresholds of the SU are*

$$\gamma_0^s = \beta \left(\log K - \log\log K + \log\log\log K\right) \tag{3.16}$$

and

$$\gamma_1^s = \beta \left[\log K - \log \log K - \log \log \log K - \log \log \log \log K - \log (\alpha \beta) \right],$$
(3.17)

where $\alpha = \frac{P^p \bar{g}_{ps}}{\bar{g}_{ss} N_0}$ *and* $\beta = \bar{g}_{ss}$. *The asymptotically optimal power allocation of the SU is given as follows.*

(i) *If* $P_{th}^s \leq \Psi(r_{RL}^p)$, *it has*

$$P_0^s = \left[\frac{P_{th}^s + \left[1 - O\left(\frac{1}{K}\right) \right] O\left(\frac{1}{\log K}\right) G^p N_0}{1 - O\left(\frac{1}{K}\right)} \right]^+,$$
(3.18)

$$P_1^s = \left[\frac{P_{th}^s - \left[1 - O\left(\frac{1}{K}\right) \right] O\left(\frac{1}{\log K}\right) F^p N_0}{1 - O\left(\frac{1}{K}\right)} \right]^+,$$
(3.19)

(ii) *if* $P_{th}^s > \Psi(r_{RL}^p)$, *it has*

$$P_0^s = \left[\frac{P_{th}^s}{F^p \left[1 - O\left(\frac{1}{K}\right) \right]} - \frac{\Theta(1) G^p \bar{g}_{pp} N_0 r_{RL}^p}{F^p \gamma^p \bar{g}_{sp} \left[1 - O\left(\frac{1}{K}\right) - r_{RL}^p \right]} \right]^+,$$
(3.20)

$$P_1^s = \left[\frac{\bar{g}_{pp} N_0 r_{RL}^p}{\gamma^p \bar{g}_{sp} \left[1 - O\left(\frac{1}{K}\right) - r_{RL}^p \right]} \right]^+,$$
(3.21)

where

$$\Psi(r_{RL}^p) = O\left(\frac{1}{\log K}\right) \left[1 - O\left(\frac{1}{K}\right) \right] F^p N_0 + \frac{\left[1 - O\left(\frac{1}{K}\right) \right] \bar{g}_{pp} N_0 r_{RL}^p}{\gamma^p \bar{g}_{sp} \left[1 - O\left(\frac{1}{K}\right) - r_{RL}^p \right]}.$$
(3.22)

Proof See Appendix A and Appendix B in [2].

Remark 3.1 The power allocation in *Theorem 3.1* is classified in terms of the relation between P_{th}^s and $\Psi(r_{RL}^p)$, which is further illustrated in Fig. 3.2. In *P3.3*, the objective function $\bar{R}^s$ is an increasing function of both P_0^s and P_1^s as shown in Fig. 3.2 a. We project the 3-D information in Fig. 3.2 a onto a 2-D plane by joining the points with equal value. The gray-scaled 2-D contour plots of $\bar{R}^s$ are shown in Fig. 3.2 b, c and d. The increase of $\bar{R}^s$ is represented by lighter contour curve and the increasing direction is pointed by the dashed arrow. Denote the secondary power constraint in (3.14) and the primary rate loss constraint in (3.15) with the equality signs by PC and RLC, respectively. The feasible regions are the shaded area enclosed by PC and RLC. For large K, (3.22) becomes $\Psi(r_{RL}^p) \approx \frac{\bar{g}_{pp} N_0 r_{RL}^p}{\gamma^p \bar{g}_{sp} (1 - r_{RL}^p)}$. The optimal power allocation is discussed in the following three cases.

• *Case 1*: If $P_{th}^s < \Psi(r_{RL}^p)$ as shown in Fig. 3.2 b, the maximum $\bar{R}^s$ is achieved at

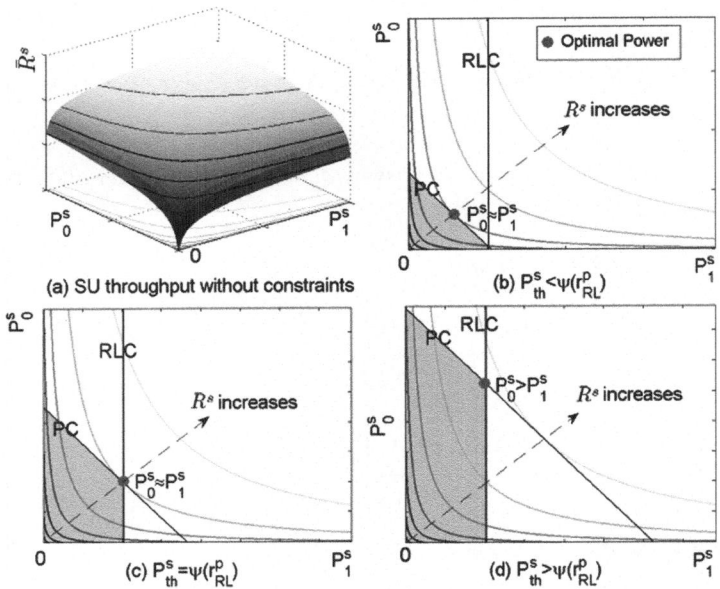

Fig. 3.2 3-D plot and contour plots of secondary throughput: **a** $\bar{R}^s$ vs P_0^s and P_1^s. **b** Contour plot of $\bar{R}^s$ for $P_{th}^s < \Psi\left(r_{RL}^p\right)$. **c** Contour plot of $\bar{R}^s$ for $P_{th}^s = \Psi\left(r_{RL}^p\right)$. **d** Contour plot of $\bar{R}^s$ for $P_{th}^s > \Psi\left(r_{RL}^p\right)$

the point where PC is tangent to the contour curve. With tight PC (small P_{th}^s), the transmit power of the SU is small, which does not violate the primary RLC. From (3.18) and (3.19), we have $P_0^s \approx P_1^s \approx P_{th}^s$ for large K. In this case, the power allocation of the SU is regardless of the PU.

• *Case 2*: If $P_{th}^s = \Psi(r_{RL}^p)$ as shown in Fig. 3.2 c, $\bar{R}^s$ obtains its maximum value at the cross point of PC and RLC which is also the tangent point as in Case 1. Similarly, it still has $P_0^s \approx P_1^s \approx P_{th}^s$ for large K.

• *Case 3*: If $P_{th}^s > \Psi(r_{RL}^p)$ as shown in Fig. 3.2d, the maximum $\bar{R}^s$ is found at the cross point of PC and RLC. With loose PC (large P_{th}^s), the increase of the secondary power is more vulnerable to RLC than PC. Hence, RLC dominates $P3.3$. From (3.20) and (3.21), we have $P_0^s \approx \left[\frac{P_{th}^s}{F^p} - \frac{G^p \bar{g}_{pp} N_0 r_{RL}^p}{F^p \gamma^p \bar{g}_{sp}(1 - r_{RL}^p)}\right]^+$ and $P_1^s \approx \frac{\bar{g}_{pp} N_0 r_{RL}^p}{\gamma^p \bar{g}_{sp}(1 - r_{RL}^p)}$ for large K. By further observation, it has $P_0^s > P_1^s$. To protect the PU, the ST uses high power and low power if the primary feedback is "0" and "1", respectively.

3.4.2   Throughput Scaling Law

With the asymptotically optimal $\{\gamma_0^s, \gamma_1^s, P_0^s, P_1^s\}$ given in *Theorem 3.1*, the max-imized secondary throughput scales double logarithmically as the increase of the number of SRs.

Theorem 3.2 *Consider a K-user downlink cognitive network with 1-bit feedback from each user. With the asymptotically optimal thresholds and power allocation for large K, the average secondary throughput scales as*

$$\bar{R}^s = \log\log K + O(1). \tag{3.23}$$

Proof See Appendix C in [2].

Remark 3.2 For the proposed 1-bit feedback scheme, the throughput scaling law of the SU is the same as the cognitive broadcast network with full channel quality information (CQI) [7]. Moreover, it also follows the same rule as that of the stand-alone broadcast network with full CQI [8] and with 1-bit feedback [9–12].

3.5 Numerical Results

In this section, we show the numerical results of the optimal thresholds, power allocation and throughput scaling law of the SU. The mean values of channel power gains are given as $\bar{g}_{pp} = 1, \bar{g}_{ps} = 0.6, \bar{g}_{sp} = 0.4$ and $\bar{g}_{ss} = 5$. Given the primary power constraint $P_{th}^p = 10$ dB, the corresponding optimal primary power and threshold in $P3.1$ are $P^p = 18.0748$ and $\gamma^p = 0.5919$, respectively. Substituting the optimal P^p and γ^p into $P3.3$, we obtain the optimal $\gamma_0^s, \gamma_1^s, P_0^s$ and P_1^s.

3.5.1 Thresholds

In Fig. 3.3, it is illustrated that the numerically optimal secondary thresholds for $P3.3$ grow almost logarithmically as the increase of the number of the SRs. We also notice that the SU uses a higher threshold for the primary feedback "0" than for the primary feedback "1", i.e., $\gamma_0^s > \gamma_1^s$. Figure 3.4 shows the asymptotically optimal thresholds in *Theorem 3.1* are approaching the numerically optimal thresholds for large K.

3.5.2 Power Allocation

Figure 3.5 discusses the relation between the numerically optimal P_0^s and P_1^s. For low secondary power constraint, i.e., $P_{th}^s \leq \Psi(r_{RL}^p)$, it has $P_0^s \approx P_1^s$ since the primary RLC is not activated. For high secondary power constraint, i.e., $P_{th}^s > \Psi\left(r_{RL}^p\right)$, it has $P_0^s > P_1^s$ due to the activation of the primary RLC. Since the ST has to restrict the power if the PU is in operation, it is beneficial to transmit with higher power if the primary feedback is "0" and lower power if the primary feedback is "1". Moreover,

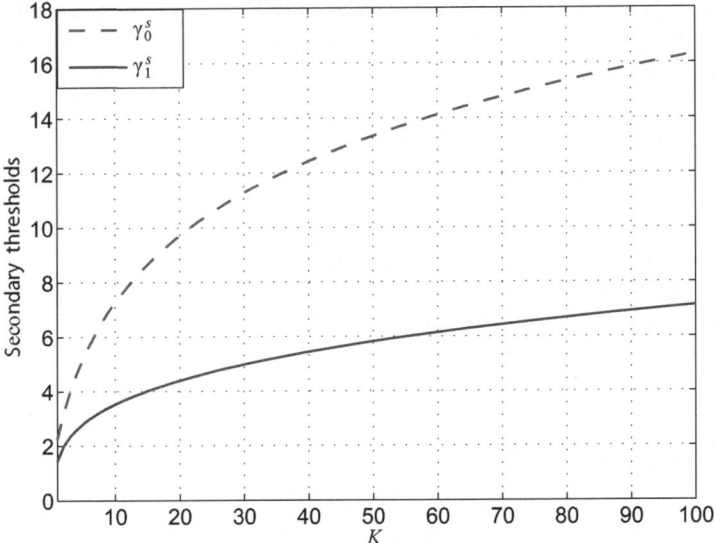

Fig. 3.3 Numerically optimal thresholds for the *SU* for small K ($K = 1$ to 10^2, $P_{th}^s = 5$ dB and $r_{RL}^p = 0.1$)

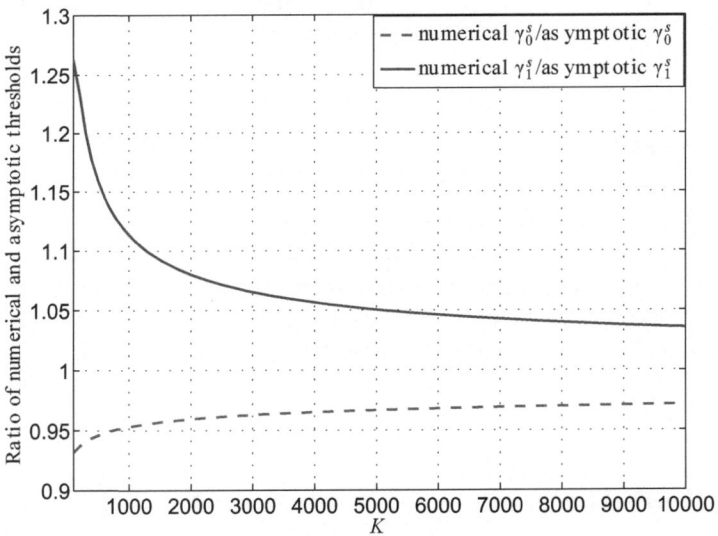

Fig. 3.4 The ratio of the numerically and asymptotically optimal thresholds of the *SU* for large K ($K = 10^2$ to 10^4, $P_{th}^s = 5$ dB and $r_{RL}^p = 0.1$)

the ratio of P_0^s / P_1^s increases dramatically as the increase of P_{th}^s. It is because the ST can allocate as much power as possible when the PU is already in outage (the primary feedback is "0"). Furthermore, given a fixed P_{th}^s, the ratio of P_0^s / P_1^s increases when

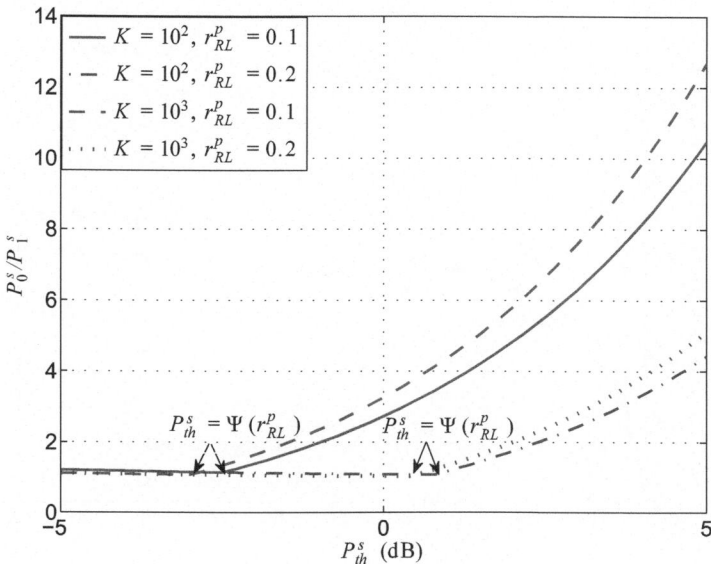

Fig. 3.5 Optimal power allocation of the SU with different power constraints

the primary RLC is tighter (smaller r_{RL}^p). The above phenomenon further validates the power allocation in *Theorem 3.1* and *Remark 3.1*.

3.5.3 Throughput

Figure 3.6 shows that the secondary throughput grows almost double logarithmically as the increase of K. The larger secondary power constraint or looser primary RLC results in better secondary throughput. In Fig. 3.7, we plot the maximized secondary throughput against $\log \log K$ for large K. Let $x = \log \log K$ and $y = \bar{R}^s$, (3.23) is rewritten as $y = x + O(1)$ which is a line with a slope of 1 and a y-intercept of $O(1)$. The parameters such as P_{th}^s and r_{RL}^p are included in the $O(1)$ term. For the curves with 1-bit feedback in Fig. 3.7, the slope is almost 1 and the y-intercepts vary with different P_{th}^s and r_{RL}^p. This result validates the throughput scaling law in *Theorem 3.2*. We also plot the reference curves with full CQI where the perfect information of the effective channel gains $\tilde{g}_{ss}^k$ of all the SRs is available at the ST. The throughput scaling law with full CQI also scales as $\log \log K$ which is the same as that with 1-bit feedback.

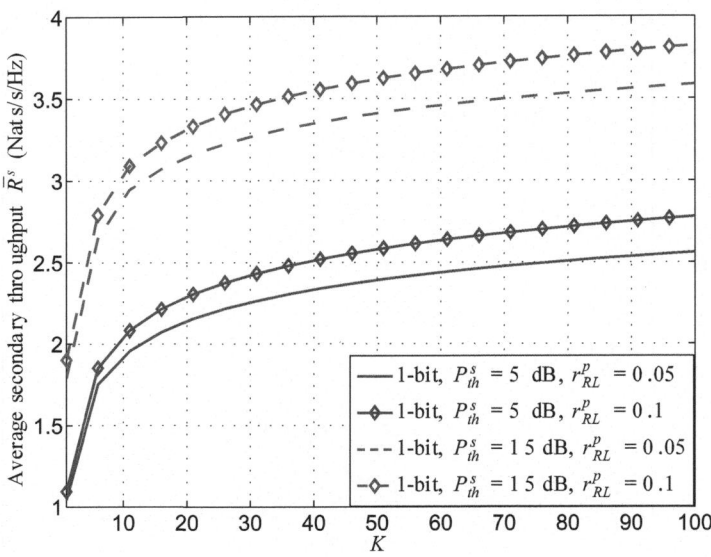

Fig. 3.6 Secondary throughput for small K ($K \in [1, 10^2]$)

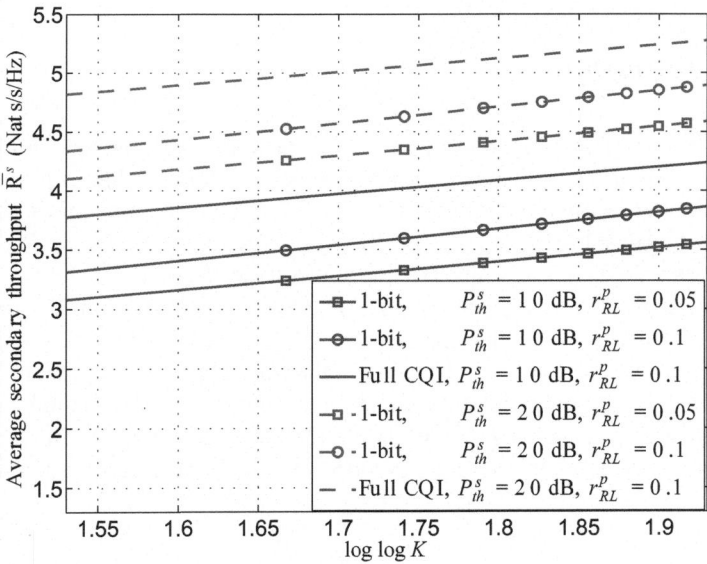

Fig. 3.7 The throughput scaling law of the SU for large K ($K \in [10^2, 10^3]$)

3.6 Summary

In this chapter, we discussed spectrum sharing with 1-bit feedback in a downlink cognitive network. The PR and K SRs each sends 1-bit feedback to its transmitter, where feedback "1" indicates the channel is in a good condition. With the overheard primary feedback and received secondary feedback, the ST schedules one of the best SR with secondary feedback "1" for transmission and adapts its power and rate to both the primary and secondary feedback. The secondary quantization thresholds and power allocation were jointly optimized and the asymptotically optimal solutions were given analytically. The results show the SU should adopt a large secondary threshold if the primary feedback is "0" and a small threshold if the primary feedback is "1". With tight secondary power constraint, the secondary power allocation does not depend much on the primary feedback. With tight primary rate loss constraint, the SU uses high power if it knows the potential primary outage event from the primary feedback "0" and low power otherwise. With the asymptotically optimal resource allocation, the secondary throughput for the proposed 1-bit feedback scheme scales as $\log \log K + O(1)$ which is the same as that with full CQI.

References

1. P. Viswanath, D. Tse, and R. Laroia, "Opportunistic beamforming using dumb antennas," *IEEE Trans. Inf. Theory*, vol. 48, no. 6, pp. 1277–1294, June 2002.
2. Z. Wang and W. Zhang, "Exploiting multiuser diversity with 1-bit feedback for spectrum sharing," *IEEE Trans. Commun.*, vol. 62, pp. 29–40, Jan. 2014.
3. T. T. Kim and M. Skoglund, "On the expected rate of slowly fading channels with quantized side information," *IEEE Trans. Commun.*, vol. 55, pp. 820–829, Apr. 2007.
4. R. Zhang, "Optimal power control over fading cognitive radio channel by exploiting primary user CSI," in *Proc. IEEE Global Commun. Conf. (GLOBECOM)*, New Orleans, USA, Nov. 2008.
5. S. Boyd and L. Vandenberghe, *Convex Optimization*, Cambridge, U.K. Cambridge Univ. Press, 2004.
6. T. H. Cormen, C. E. Leiserson, R. L. Rivest, and C. Stein, *Introduction to Algorithms (3rd Edition)*, The MIT Press, 2009.
7. Y. Li and A. Nosratinia, "Capacity limits of multiuser multiantenna cognitive networks," *IEEE Trans. Inf. Theory*, vol. 7, pp. 4493–4508, Jul. 2012.
8. M. Sharif and B. Hassibi, "On the capacity of MIMO broadcast channels with partial side information," *IEEE Trans. Inf. Theory*, vol. 51, pp. 506–522, Feb. 2005.
9. S. Sanayei and A. Nosratinia, "Exploiting multiuser diversity with only 1-bit feedback," *Proc. IEEE Wireless Commun. and Networking Conf. (WCNC)*, New Orleans, LA, USA, Mar. 2005.
10. S. Sanayei and A. Nosratinia, "Opportunistic downlink transmission with limited feedback," *IEEE Trans. Inf. Theory*, vol. 53, pp. 4363–4372, Nov. 2007.
11. J. Diaz, O. Simeone, and Y. Bar-Ness, "Sum-rate of MIMO broadcast channels with one bit feedback," in *Proc. IEEE Int. Symp. Inf. Theory (ISIT 2006)*, Seattle, USA, July 9–14, 2006.
12. B. Niu, O. Simeone, O. Somekh, and A. M. Haimovich, "Ergodic and outage performance of fading broadcast channels with 1-bit feedback," *IEEE Trans. Veh. Technol.*, vol. 59, pp. 1282–1293, Mar. 2010.

Chapter 4
Cognitive Ad Hoc Network with Limited Feedback

Abstract In this chapter, we study the opportunistic spectrum sharing in ad hoc networks. Both primary and secondary transmitters are modeled as Poisson point processes. The primary and secondary receivers each sends 1-bit feedback of local channel gain to their corresponding transmitters. The primary transmitters are active if their local primary channel gains are above a certain threshold. In the first scheme, the secondary transmitters transmit if their local channel gains are above the required threshold. In the second scheme, the secondary transmitters transmit if their local channel gains are above the threshold and they are outside the primary exclusive regions of the active primary receivers. Using stochastic geometry theory, the analytical solutions of the optimal secondary node density are derived by maximizing the average secondary throughput per unit area under the reliability constraints of both primary and secondary users. With tight secondary constraint, the scheme without primary exclusive region is more beneficial. With tight primary constraint, the scheme with primary exclusive regions is recommended.

Keywords Cognitive radio · Spectrum sharing · Opportunistic Aloha · Primary exclusive region · Poisson point process · Stochastic geometry theory

4.1 Introduction

In Chaps. 2 and 3, we studied the opportunistic spectrum sharing in the point-to-point and broadcast cognitive network. In this chapter, we extend the discussion into the ad hoc network [1, 2]. The detection and utilization of the temporal and spatial spectrum holes are discussed in a large random network, with the aid of limited channel feedback. Stochastic geometry theory [3] is used to model the spatial patterns of the user positions and provide tractable network performance in the large scale network [4–10].

As discussed, interference management is one of the challenging issues for spectrum sharing. Instead of receiving interference from a single secondary transmitter

---

© [2014] IEEE. Reprinted, with permission, from [Z. Wang and W. Zhang, "Opportunistic spectrum sharing with limited feedback in Poisson cognitive radio networks," *IEEE Trans. Wireless Commun.*, vol. 13, no. 12, pp. 7098–7109, Dec. 2014.

© The Author(s) 2015
Z. Wang, W. Zhang, *Opportunistic Spectrum Sharing in Cognitive Radio Networks,*
SpringerBriefs in Electrical and Computer Engineering, DOI 10.1007/978-3-319-15542-5_4

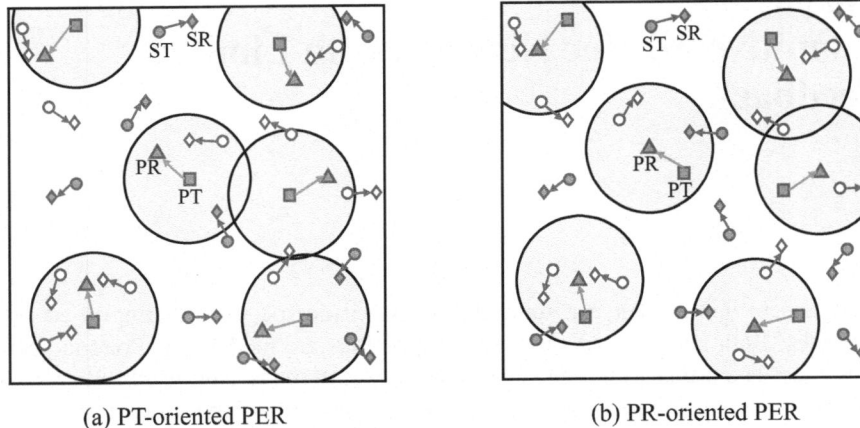

    (a) PT-oriented PER   (b) PR-oriented PER

Fig. 4.1 PT-oriented PER and PR-oriented PER. The PERs are represented by the *big shaded circles*. The *PT*s, *PR*s, *ST*s and *SR*s are denoted by *squares*, *triangles*, *circles* and *diamonds*, respectively. The active and inactive *ST*s are represented by *solid and hollow circles*, respectively

(ST), in ad hoc network, the primary receiver (PR) may suffer the aggregate interference from a large number of concurrent STs. The interference management are classified into two categories.

- In the first category, the instantaneous performance of the primary user (PU) is protected by the primary exclusive regions (PERs) which are the circular regions centered at the primary transmitters (PTs) or PRs. The STs are not allowed to transmit if they are inside the PERs. PT-oriented PER [11–13] and PR-oriented PER [13–15] are shown in Fig. 4.1a and b, respectively. If PRs are passive, PT-oriented PER is adopted though the active STs may still cause harmful interference to the nearby PRs. If PRs are able to broadcast some beacon signals [13], PR-oriented PER is more efficient in PU protection.
- In the second category, PER is not applicable due to the lack of primary location information at the SU. In this case, the secondary node density or transmit power should be controlled so as not to violate the tolerable quality of service (QoS) degradation of the PU, e.g., average throughput degradation constraint [16, 17] and outage probability constraint [18, 19].

Due to the channel fluctuation, the secondary user (SU) may suffer from low throughput if its local channel is in deep fading and waste transmission opportunity if it is inside a PER of a deep faded PU. Opportunistic Aloha protocol [20, 4] eliminates the effects of deep fading channels by enabling only the transmitters with good local channels to be active. With less concurrent transmitters, the aggregate interference at each node is also reduced. In this chapter, we exploit the channel opportunities in ad hoc cognitive network via limited feedback. Two spectrum sharing schemes are discussed, without and with PER, respectively.

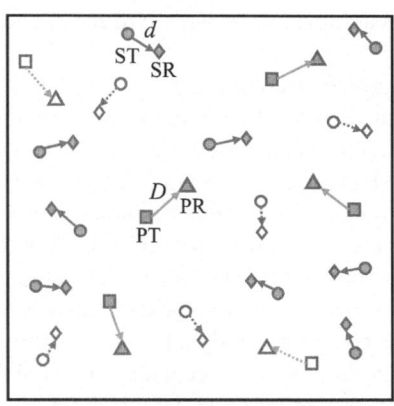

(a) Scheme I: without PER

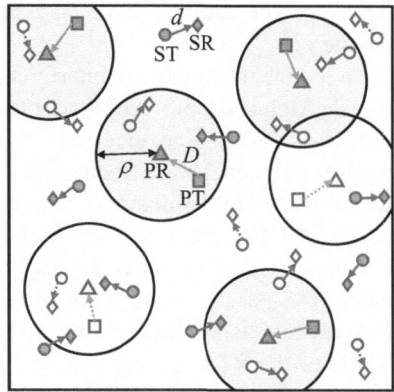

(b) Scheme II: with PER

Fig. 4.2 System model of the ad-hoc spectrum sharing network. The *PT*s and *PR*s are represented by the *squares* and *triangles*, respectively, where the distance between each pair of *PT* and *PR* is *D*. The *ST*s and *SR*s are denoted by *circles* and *diamonds*, respectively, where the distance between each pair of *ST* and *SR* is *d*. The *solid and dashed arrows* denote the channel gain is above and below the threshold, respectively. The active and inactive users are denoted by *filled* and *hollow icons*, respectively. The active and inactive PERs are represented by the *shaded and hollow big circles* with radius ρ

The rest of the chapter is organized as follows. In Sect. 4.2, system model is introduced. In Sect. 4.3, we derive the optimal node density of the PU that maximizes the average primary throughput. In Sect. 4.4, the opportunistic spectrum sharing problem is presented. In Sects. 4.5 and 4.6, the optimal node density of the SU that maximizes the average secondary throughput is derived for the schemes without and with PER, respectively. In Sect. 4.7, the numerical results are shown. Finally, the summary is drawn in Sect. 4.8.

4.2 System Model

As shown in Fig. 4.2, a secondary ad hoc network shares the same frequency band with a primary ad hoc network in a $\mathbb{R}^2$ plane. Primary transmitters (PT_j) and secondary transmitters (ST_i) follow two independent homogeneous Poisson point processes $\Phi_p = \{X_j\}$ and $\Phi_s = \{Y_i\}$ with density λ_p and λ_s, respectively. Primary receivers (PR_j) and secondary receivers (SR_i) are located at D and d distances away from their corresponding transmitters PT_j and ST_i in random directions, respectively. Denote $h_{pp}^{a,b}$, $h_{ps}^{a,b}$, $h_{sp}^{a,b}$ and $h_{ss}^{a,b}$ the channel fading gains between any pair of PT_a-PR_b, PT_a-SR_b, ST_a-PR_b and ST_a-SR_b, respectively, where a is the index of transmitter and b is the receiver index. The channel fading gains follow i.i.d. exponential distribution with unit mean. Denote $\ell(\cdot) = |\cdot|^{-\alpha}$ the path loss function, where α is the path loss exponent. The additive white Gaussian noise is assumed to be with zero mean and variance N_0.

To reduce the aggregate interference and eliminate deep fading channels, opportunistic Aloha [20] protocol is adopted by both the PUs and SUs. Only the users with good local channels are elected to transmit. For local channel evaluation, the sender-receiver pairs exchange Request-To-Send (RTS) and Clear-To-Send (CTS) packets in CDMA or TDMA manners, which separates neighboring channel estimation processes [20]. For example, each Aloha slot is divided into three parts: RTS, CTS and data transmission sub-slots. During the RTS sub-slot, the PTs and STs send CDMA coded RTS packets to their corresponding receivers. The PRs and SRs receive the RTS packets using parallel matched filters and compare the signal-to-noise-ratios (SNR) of the RTS packets with the primary threshold θ_p and secondary threshold θ_s, respectively. If the SNR of the received RTS packet is above the threshold, the receiver sends the CTS packet with indicator "1" back to its transmitter using the same CDMA code as the RTS packet in the following CTS sub-slot. Otherwise, the feedback indicator is "0". With channel reciprocity, the reception of the CTS packets is similar to that of the RTS packets.

4.2.1   PU Transmission Scheme

PR_j feeds back "1" or "0" to PT_j if the local SNR is above or below θ_p, respectively. The feedback indicator of PR_j is given by

$$b_p^j = \mathbb{1}\left(h_{pp}^{j,j} \geq D^\alpha N_0 \theta_p / \sigma_p\right) = \begin{cases} 1, & \text{if } h_{pp}^{j,j} \geq D^\alpha N_0 \theta_p / \sigma_p \\ 0, & \text{if } h_{pp}^{j,j} < D^\alpha N_0 \theta_p / \sigma_p. \end{cases} \tag{4.1}$$

The transmit power of PT_j is

$$P_{PT_j} = \begin{cases} \sigma_p, & \text{if } b_p^j = 1 \\ 0, & \text{if } b_p^j = 0. \end{cases} \tag{4.2}$$

4.2.2   SU Transmission Schemes

Similarly, SR_i sends feedback "1" or "0" to ST_j if its local SNR is above or below θ_s, respectively. The feedback indicator of SR_i is given by

$$b_s^i = \mathbb{1}\left(h_{ss}^{i,i} \geq d^\alpha N_0 \theta_s / \sigma_s\right) = \begin{cases} 1, & \text{if } h_{ss}^{i,i} \geq d^\alpha N_0 \theta_s / \sigma_s \\ 0, & \text{if } h_{ss}^{i,i} < d^\alpha N_0 \theta_s / \sigma_s. \end{cases} \tag{4.3}$$

We consider two opportunistic spectrum sharing schemes for the SU, as shown in Fig. 4.2a and b, respectively.

4.2.2.1   SU Transmission Scheme I (Without PERs)

In Scheme I, PER is not applicable due to the lack of location information of the PRs at the STs. ST_i transmits if the local feedback from SR_i is "1". The transmit power of ST_i in Scheme I is given by

$$P_{ST_i}^{(I)} = \begin{cases} \sigma_s, & \text{if } b_s^i = 1 \\ 0, & \text{if } b_s^i = 0. \end{cases} \qquad (4.4)$$

4.2.2.2   SU Transmission Scheme II (with PERs)

In Scheme II, the STs are assumed to have the location information of the PRs, e.g., by accessing the primary geo-location database [21]. We adopt PR-oriented PER to control the instantaneous interference from the STs to PRs. Each PR_j is protected by a PER which is a circular restriction region centered at PR_j with radius ρ ($\rho > 0$). PER_j is regarded as active or inactive if the corresponding feedback from the PR_j is "1" or "0", respectively. ST_i is regarded as a potential active ST if its receives feedback "1" from SR_i. We assume that STs inside the PERs know if this PER is active or not based on the primary feedback. The potential active ST defers the transmission if it is inside the active PERs, and transmits if it is inside the inactive PERs or outside the PERs. To sum up, ST_i is active if *both* conditions are satisfied: (1) ST_i receives local feedback "1" from SR_i; (2) ST_i is outside the *active* PERs. The transmit power of ST_i in Scheme II is given by

$$P_{ST_i}^{(II)} = \begin{cases} \sigma_s, & \text{if } b_s^i = 1 \text{ and } ST_i \text{ is outside active PERs} \\ 0, & \text{otherwise.} \end{cases}$$

4.3   Optimal PU Density

In this section, the optimal primary density λ_p in the PU network is designed. Since the PUs should be oblivious to the existence of the SUs, λ_p is designed without the consideration of the SU network.

Consider a typical pair of primary transmitter PT_0 and receiver PR_0 with PR_0 at the origin. Denote $h_{pp}^{0,0}$ and $h_{pp}^{j,0}$ the channel power gains of PT_0-PR_0 and PT_j-PR_0, respectively. The Euclidean distance between PT_j and PR_0 is $|x_j|$. PR_0 receives interference from all other active PTs with $h_{pp}^{j,j} \geq D^\alpha N_0 \theta_p / \sigma_p$. Given an active PT_0, the signal-to-interference-and-noise-ratio (SINR) at PR_0 is

$$\text{SINR}_p = \frac{\sigma_p h_{pp}^{0,0} D^{-\alpha}}{\sum\limits_{x_j \in \Phi_p} I_{pp}^{j,0} + N_0},$$

where $I_{pp}^{j,0} = \sigma_p h_{pp}^{j,0} |x_j|^{-\alpha} \mathbb{1}\left(h_{pp}^{j,j} \geq D^\alpha N_0 \theta_p / \sigma_p\right)$ is the interference from PT_j to PR_0. The transmission is successful if PT_0 transmits and the received $SINR_p$ at PR_0 is above a predefined threshold θ_p. The success probability of the PU is given by

$$p_p = \Pr\left(SINR_p \geq \theta_p, h_{pp}^{0,0} \geq D^\alpha N_0 \theta_p / \sigma_p\right). \qquad (4.5)$$

Note, if PT_0 is inactive, i.e., $h_{pp}^{0,0} < D^\alpha N_0 \theta_p / \sigma_p$, we consider it as an outage event. Otherwise, if the primary channel never exceeds the target threshold, we have the $p_p \to 1$, which is unreasonable. Based on the derivation in Appendix A, the closed-form expression is given by

$$p_p = \exp\left[-\lambda_p CD^2 \left(\theta_p\right)^{\frac{2}{\alpha}} \exp\left(-\frac{D^\alpha N_0 \theta_p}{\sigma_p}\right)\right] \exp\left(-\frac{D^\alpha N_0 \theta_p}{\sigma_p}\right), \qquad (4.6)$$

where $C = \frac{2\pi^2}{\alpha \sin\left(\frac{2\pi}{\alpha}\right)}$.

Define area spectral efficiency (ASE) the average throughput per unit area, which is the product of the successful node density and target rate per user [4]. The successful node density of the PU is the average number of successful primary transmissions per unit area $\lambda_p p_p$. The target rate per PU is $R_p = \log\left(1 + \theta_p\right)$. Note $\log(\cdot)$ is natural logarithm function throughout this chapter. The ASE of the PU network at the absence of the SU network is given by

$$\bar{R}_p = \lambda_p p_p \log\left(1 + \theta_p\right). \qquad (4.7)$$

As observed, p_p is a decreasing function of λ_p, and $\bar{R}_p$ is unimodal in λ_p. More concurrent transmissions cause more interference, which thereby reduces the successful probability. To obtain the optimal primary density, we maximize the primary ASE while guaranteeing the outage probability of the PU is below the constraint ε_p.

$$P4.1 : \max_{\lambda_p} \bar{R}_p \qquad (4.8)$$

$$\text{s.t.} \quad 1 - p_p \leq \varepsilon_p. \qquad (4.9)$$

Applying Karush-Kuhn-Tucker (KKT) conditions [22] to $P4.1$, the optimal primary density λ_p^* is

$$\lambda_p^* = \frac{\left[-\frac{D^\alpha N_0 \theta_p}{\sigma_p} - \log\left(1 - \varepsilon_p\right)\right]^+}{CD^2(\theta_p)^{\frac{2}{\alpha}} \exp\left(-\frac{D^\alpha N_0 \theta_p}{\sigma_p}\right)}, \qquad (4.10)$$

where $(\cdot)^+ = \max(0, \cdot)$. Substituting (4.10) into (4.8), the maximized ASE of the PU is

$$\bar{R}_p = \frac{\left[-\frac{D^\alpha N_0 \theta_p}{\sigma_p} - \log\left(1 - \varepsilon_p\right)\right]^+}{CD^2(\theta_p)^{\frac{2}{\alpha}} \exp\left(-\frac{D^\alpha N_0 \theta_p}{\sigma_p}\right)} (1 - \varepsilon_p) \log\left(1 + \theta_p\right). \qquad (4.11)$$

Remark 4.1 If $0 \leq \varepsilon_p \leq \kappa_p$, we have $\lambda_p^* = 0$ since $1 - p_p > \varepsilon_p$ always holds, where $\kappa_p = 1 - \exp\left(-D^\alpha N_0 \theta_p / \sigma_p\right)$. The primary outage constraint is too tight for the PU. If $\kappa_p < \varepsilon_p \ll 1$, we have $\log\left(1 - \varepsilon_p\right) \approx -\varepsilon_p$ and $1 - \varepsilon_p \approx 1$, with which, the approximation of λ_p^* and $\bar{R}_p$ in (4.10) and (4.11) are almost linear in ε_p.

4.4 Spectrum Sharing Optimization

The optimal primary density λ_p^* was obtained in the previous section without considering the existence of the SUs. When the SUs shares the spectrum with the PUs, the PUs may suffer from some throughput degradation. The relative degradation of the primary ASE is defined as primary efficiency loss ratio, i.e.,

$$\delta = \frac{\bar{R}_p - \bar{R}_p'}{\bar{R}_p}, \tag{4.12}$$

where $\bar{R}_p$ and $\bar{R}_p'$ are the primary ASE at the absence and presence of the SUs, respectively. $\bar{R}_p$ has been given in (4.11) and $\bar{R}_p'$ for Scheme I and Scheme II are derived in Sects. 4.5.1 and 4.6.1, respectively. To control the interference from the SUs to PUs, δ should be restricted below the primary efficiency loss constraint (ELC) r_{th}, i.e., $\delta \leq r_{th}$, where r_{th} is the maximum allowable primary efficiency loss ratio.

To obtain the optimal secondary density λ_s, the secondary ASE $\bar{R}_s$ is maximized under the primary ELC r_{th} and secondary outage constraint ε_s.

$$P4.2 : \max_{\lambda_s} \bar{R}_s \tag{4.13}$$

$$\text{s.t.} \quad 1 - p_s \leq \varepsilon_s \tag{4.14}$$

$$\delta \leq r_{th}, \tag{4.15}$$

where p_s is the secondary success probability.

In Sects. 4.5 and 4.6, we design the optimal secondary density for the opportunistic spectrum sharing schemes without PERs and with PERs, respectively. The superscripts of (I) and (II) are used to distinguish the parameters in Scheme I and Scheme II when necessary.

4.5 Scheme I: Without Primary Exclusive Regions

In this scheme, neither location nor feedback information of the PUs is available at the SUs. Since PER is not applicable, the SUs protect the average PU performance by the primary ELC. ST_i is elected to transmit with power σ_s if it receives feedback "1" from SR_i. In the following discussions, we firstly determine the expressions of $\bar{R}_s$, δ and p_s for Scheme I, and then derive the optimal secondary density that maximizes the secondary ASE in $P4.2$.

4.5.1 PU Efficiency Loss Ratio

Consider a typical PR_0 at the origin. With the existence of the SUs, PR_0 receives the aggregate interference from all other active PTs and all active STs. Denote the interference channel power gain and Euclidean distance between ST_i and the PR_0 by $h_{sp}^{i,0}$ and $|y_i|$, respectively. Given PT_0 is active, the SINR at the PR_0 is given by

$$\text{SINR}_p^{'(\text{I})} = \frac{\sigma_p h_{pp}^{0,0} D^{-\alpha}}{\sum\limits_{x_j \in \Phi_p} I_{pp}^{j,0} + \sum\limits_{y_i \in \Phi_s} I_{sp}^{i,0\,(\text{I})} + N_0}, \tag{4.16}$$

where

$$I_{pp}^{j,0} = \sigma_p h_{pp}^{j,0} |x_j|^{-\alpha} \mathbf{1}\left(h_{pp}^{j,j} \geq D^\alpha N_0 \theta_p / \sigma_p\right) \tag{4.17}$$

and

$$I_{sp}^{i,0\,(\text{I})} = \sigma_s h_{sp}^{i,0} |y_i|^{-\alpha} \mathbf{1}\left(h_{ss}^{i,i} \geq d^\alpha N_0 \theta_s / \sigma_s\right) \tag{4.18}$$

are the interference from PT_j and ST_i to PR_0, respectively. The superscript (I) stands for Scheme I. The success probability of the PUs in the presence of the SUs is

$$p_p^{'(\text{I})} = \Pr\left(\text{SINR}_p^{'(\text{I})} \geq \theta_p, h_{pp}^{0,0} \geq D^\alpha N_0 \theta_p / \sigma_p\right). \tag{4.19}$$

By further derivations, we have

$$p_p^{'(\text{I})} = \exp\left(-\frac{D^\alpha N_0 \theta_p}{\sigma_p}\right) \exp\left[-\lambda_p^* C D^2 \left(\theta_p\right)^{\frac{2}{\alpha}} \exp\left(-\frac{D^\alpha N_0 \theta_p}{\sigma_p}\right)\right]$$

$$\times \exp\left[-\lambda_s C D^2 \left(\frac{\sigma_s \theta_p}{\sigma_p}\right)^{\frac{2}{\alpha}} \exp\left(-\frac{d^\alpha N_0 \theta_s}{\sigma_s}\right)\right]. \tag{4.20}$$

Note, PUs still use λ_p^* in (4.10) when the SUs are introduced since the PUs are not aware of the existence of the SUs.

The primary ASE in the presence of SUs is given by

$$\bar{R}_p^{'(\text{I})} = \lambda_p^* p_p^{'(\text{I})} \log\left(1 + \theta_p\right). \tag{4.21}$$

Based on (4.11) and (4.21), the primary efficiency loss ratio is given by

$$\delta^{(\text{I})} = \frac{\bar{R}_p - \bar{R}_p^{'(\text{I})}}{\bar{R}_p} = 1 - \exp\left[-\lambda_s C D^2 \left(\frac{\sigma_s \theta_p}{\sigma_p}\right)^{\frac{2}{\alpha}} \exp\left(-\frac{d^\alpha N_0 \theta_s}{\sigma_s}\right)\right]. \tag{4.22}$$

4.5.2 SU Area Spectral Efficiency

Consider a typical secondary transmitter ST_0 and receiver SR_0 with SR_0 at the origin. SR_0 receives the aggregate interference from all other active STs and all active PTs. Denote the channel gains of ST_0-SR_0, ST_i-SR_0 and PT_j-SR_0 by $h_{ss}^{0,0}$, $h_{ss}^{i,0}$ and $h_{ps}^{j,0}$, respectively. Given ST_0 transmits, the SINR at SR_0 is given by

$$\text{SINR}_s^{(\text{I})} = \frac{\sigma_s h_{ss}^{0,0} d^{-\alpha}}{\sum\limits_{x_j \in \Phi_p} I_{ps}^{j,0\ (\text{I})} + \sum\limits_{y_i \in \Phi_s} I_{ss}^{i,0\ (\text{I})} + N_0}, \tag{4.23}$$

where

$$I_{ps}^{j,0\ (\text{I})} = \sigma_p h_{ps}^{j,0} |x_j|^{-\alpha} \mathbb{1}\left(h_{pp}^{j,j} \geq D^\alpha N_0 \theta_p / \sigma_p\right) \tag{4.24}$$

and

$$I_{ss}^{i,0\ (\text{I})} = \sigma_s h_{ss}^{i,0} |y_i|^{-\alpha} \mathbb{1}\left(h_{ss}^{i,i} \geq d^\alpha N_0 \theta_s / \sigma_s\right) \tag{4.25}$$

are the interference from PT_j and ST_i to SR_0, respectively.

The transmission is successful if ST_0 transmits and the received SINR at SR_0 is above the threshold θ_s. Similar to (4.20), the secondary success probability is written as

$$\begin{aligned}
p_s^{(\text{I})} &= \Pr\left(\text{SINR}_s^{(\text{I})} \geq \theta_s, h_{ss}^{0,0} \geq d^\alpha N_0 \theta_s / \sigma_s\right) \\
&= \exp\left(-\frac{d^\alpha N_0 \theta_s}{\sigma_s}\right) \exp\left[-\lambda_p^* C d^2 \left(\frac{\sigma_p \theta_s}{\sigma_s}\right)^{\frac{2}{\alpha}} \exp\left(-\frac{D^\alpha N_0 \theta_p}{\sigma_p}\right)\right] \\
&\quad \times \exp\left[-\lambda_s C d^2 \left(\theta_s\right)^{\frac{2}{\alpha}} \exp\left(-\frac{d^\alpha N_0 \theta_s}{\sigma_s}\right)\right].
\end{aligned} \tag{4.26}$$

The secondary ASE is given by

$$\bar{R}_s^{(\text{I})} = \lambda_s p_s^{(\text{I})} \log\left(1 + \theta_s\right). \tag{4.27}$$

4.5.3 Optimal SU Density

Substituting (4.10), (4.22), (4.26) and (4.27) into $P4.2$ and solving it via KKT conditions, we obtain the optimal secondary density as follows.
(1) If $\varepsilon_s \leq \eta^{(\text{I})}(r_{th})$,

$$\lambda_s^{*(\text{I})} = \frac{\left[\dfrac{\log(1-\varepsilon_p) + \frac{D^\alpha \theta_p N_0}{\sigma_p}}{D^2 \left(\frac{\sigma_s \theta_p}{\sigma_p}\right)^{\frac{2}{\alpha}}} - \dfrac{\log(1-\varepsilon_s) + \frac{d^\alpha \theta_s N_0}{\sigma_s}}{d^2 (\theta_s)^{\frac{2}{\alpha}}}\right]^+}{C \exp\left(-\frac{d^\alpha \theta_s N_0}{\sigma_s}\right)} \tag{4.28}$$

$$\bar{R}_s^{(I)} = \left[\frac{\log\left(1 - \varepsilon_p\right) + \frac{D^\alpha \theta_p N_0}{\sigma_p}}{D^2 \left(\frac{\sigma_s \theta_p}{\sigma_p}\right)^{\frac{2}{\alpha}}} - \frac{\log\left(1 - \varepsilon_s\right) + \frac{d^\alpha \theta_s N_0}{\sigma_s}}{d^2 (\theta_s)^{\frac{2}{\alpha}}} \right]^+$$

$$\times \frac{1 - \varepsilon_s}{C \exp\left(-\frac{d^\alpha \theta_s N_0}{\sigma_s}\right)} \log\left(1 + \theta_s\right) \tag{4.29}$$

(2) If $\varepsilon_s \geq \eta^{(I)}(r_{th})$,

$$\lambda_s^{*(I)} = -\frac{\log\left(1 - r_{th}\right)}{CD^2 \left(\frac{\sigma_s \theta_p}{\sigma_p}\right)^{\frac{2}{\alpha}} \exp\left(-\frac{d^\alpha N_0 \theta_s}{\sigma_s}\right)} \tag{4.30}$$

$$\bar{R}_s^{(I)} = -\frac{\log\left(1 - r_{th}\right)}{CD^2 \left(\frac{\sigma_s \theta_p}{\sigma_p}\right)^{\frac{2}{\alpha}}} \left[\frac{(1 - \varepsilon_p)(1 - r_{th})}{\exp\left(-\frac{D^\alpha N_0 \theta_p}{\sigma_p}\right)} \right]^{\frac{(\sigma_p \theta_s)^{\frac{2}{\alpha}} d^2}{(\sigma_s \theta_p)^{\frac{2}{\alpha}} D^2}} \log\left(1 + \theta_s\right), \tag{4.31}$$

where

$$\eta^{(I)}(r_{th}) = 1 - \exp\left(-\frac{d^\alpha N_0 \theta_s}{\sigma_s}\right) \left[\frac{(1 - \varepsilon_p)(1 - r_{th})}{\exp\left(-\frac{D^\alpha N_0 \theta_p}{\sigma_p}\right)} \right]^{\frac{(\sigma_p \theta_s)^{\frac{2}{\alpha}} d^2}{(\sigma_s \theta_p)^{\frac{2}{\alpha}} D^2}}. \tag{4.32}$$

Remark 4.2 If $0 \leq \varepsilon_s \leq \kappa_s^{(I)}$, we have $\lambda_s^{*(I)} = 0$ since $1 - p_s > \varepsilon_s$ always holds, where

$$\kappa_s^{(I)} = 1 - \left[\frac{1 - \varepsilon_p}{\exp\left(-\frac{D^\alpha N_0 \theta_p}{\sigma_p}\right)} \right]^{\frac{(\sigma_p \theta_s)^{\frac{2}{\alpha}} d^2}{(\sigma_s \theta_p)^{\frac{2}{\alpha}} D^2}} \exp\left(-\frac{d^\alpha N_0 \theta_s}{\sigma_s}\right). \tag{4.33}$$

Due to the tight SU outage constraint, the optimal secondary density is set to be zero. If $\kappa_s^{(I)} < \varepsilon_s \leq \eta^{(I)}(r_{th})$, we have $\log\left(1 - \varepsilon_s\right) \approx -\varepsilon_s$ and $1 - \varepsilon_s \approx 1$, hence, $\lambda_s^{*(I)}$ and $\bar{R}_s^{(I)}$ increase almost linearly with ε_s. If $\varepsilon_s \geq \eta^{(I)}(r_{th})$, $\lambda_s^{*(I)}$ and $\bar{R}_s^{(I)}$ are constants with respect to ε_s.

4.6 Scheme II: With Primary Exclusive Regions

In this scheme, the location information of PRs is available at the STs. The instantaneous and average performance of the PUs are protected by PERs and ELC, respectively. As discussed Sect. 4.2.2, ST_i is active if it receives local feedback "1" from SR_i and it is outside the active PERs. Compared to Scheme I, the introduction of PER may bring both benefit and restriction to the SU. On the one hand, PERs provide better protection for the PUs, which may accommodate more concurrent secondary transmissions. From this perspective, $\bar{R}_s$ may increase. On the other hand, more concurrent secondary transmissions result in higher secondary outage probability, which may restrict $\bar{R}_s$. In other words, PER may reduce the primary efficiency loss ratio at the cost of higher secondary outage probability.

In the following discussions, we firstly determine the expressions of $\bar{R}_s$, δ and p_s for Scheme II and then derive the optimal secondary density that maximizes the secondary ASE in $P4.2$. Furthermore, the effect of the PER radius ρ is discussed.

4.6.1 PU Efficiency Loss Ratio

Consider a typical primary transmitter PT_0 and receiver PR_0 with PR_0 at the origin of the network. Given PT_0 is active, the received SINR at PR_0 is given by

$$\text{SINR}_p^{'(\text{II})} = \frac{\sigma_p h_{pp}^{0,0} D^{-\alpha}}{\sum\limits_{x_j \in \Phi_p} I_{pp}^{j,0} + \sum\limits_{y_i \in \Phi_s} I_{sp}^{i,0\,(\text{II})} + N_0}, \tag{4.34}$$

where

$$I_{pp}^{j,0} = \sigma_p h_{pp}^{j,0} |x_j|^{-\alpha} \mathbb{1}\left(h_{pp}^{j,j} \geq \frac{D^{\alpha} N_0 \theta_p}{\sigma_p} \right) \tag{4.35}$$

and

$$I_{sp}^{i,0\,(\text{II})} = \sigma_s h_{sp}^{i,0} |y_i|^{-\alpha} \mathbb{1}\left(h_{ss}^{i,i} \geq \frac{d^{\alpha} N_0 \theta_s}{\sigma_s} \right) \mathbb{1}\,(ST_i \text{ is outside the active PERs}) \tag{4.36}$$

are the interference from PT_j and ST_i to PR_0, respectively. The primary success probability is given by

$$p_p^{'(\text{II})} = \Pr\left(\text{SINR}_p^{'(\text{II})} \geq \theta_p, h_{pp}^{0,0} \geq \frac{D^{\alpha} N_0 \theta_p}{\sigma_p} \right). \tag{4.37}$$

Assume the interference from the PTs and STs are independent. Similar to (4.20), we have

$$p_p^{'(\text{II})} = \exp\left(-\frac{D^{\alpha} \theta_p N_0}{\sigma_p} \right) \mathbb{E}\left[\exp\left(-D^{\alpha} \theta_p \sum_{x_j \in \Phi_p} I_{pp}^{j,0} \right) \right]$$

$$\times \mathbb{E}\left[\exp\left(-D^{\alpha} \theta_p \sum_{y_i \in \Phi_s} I_{sp}^{i,0\,(\text{II})} \right) \right]. \tag{4.38}$$

The aggregate interference from all other active PTs to PR_0 has been discussed in the previous sections. Based on Appendix A, the second term in (4.38) is written as

$$
\mathbb{E}\left[\exp\left(-D^\alpha\theta_p\sum_{x_j\in\Phi_p}I_{pp}^{j,0}\right)\right]=\exp\left[-\lambda_p^*CD^2\left(\theta_p\right)^{\frac{2}{\alpha}}\exp\left(-\frac{D^\alpha\theta_pN_0}{\sigma_p}\right)\right].
$$
(4.39)

The active node density of the PUs is $\lambda_p^*\exp\left(-\frac{D^\alpha N_0\theta_p}{\sigma_p}\right)$.

We next derive the third term of (4.38). As shown in (4.26), the potential active SUs form another homogeneous PPP with density of $\lambda_s\exp\left(-\frac{d^\alpha N_0\theta_s}{\sigma_s}\right)$. Since the active STs should stay at least of distance ρ from the active PRs, they no longer follow a PPP but a Poisson hole process [3, 14]. Since the Laplace transform of Poisson hole process is not tractable, we use approximations in the following analyses. The probability of having no active STs inside the PER centered at an active PR is equivalent to the probability of having no active PRs inside the PER centered at an active ST, which is given by

$$
\Pr\left[N(\pi\rho^2)=0\right]=\exp\left[-\lambda_p^*\pi\rho^2\exp\left(-\frac{D^\alpha N_0\theta_p}{\sigma_p}\right)\right].
$$
(4.40)

Since ST_j is active if receiving feedback "1" from SR_j while being outside the active PERs, its active probability is $\exp\left(-\frac{d^\alpha N_0\theta_s}{\sigma_s}\right)\exp\left[-\lambda_p^*\pi\rho^2\exp\left(-\frac{D^\alpha N_0\theta_p}{\sigma_p}\right)\right]$. We approximate the Poisson hole process of the active STs as a homogeneous PPP with a minimum distance ρ to PR_0. The secondary density of the new homogenous PPP is $\lambda_s\exp\left(-\frac{d^\alpha N_0\theta_s}{\sigma_s}\right)\exp\left[-\lambda_p^*\pi\rho^2\exp\left(-\frac{D^\alpha N_0\theta_p}{\sigma_p}\right)\right]$. According to [8, (3.46)], we have

$$
\mathbb{E}\left[\exp\left(-D^\alpha\theta_p\sum_{x_j\in\Phi_s}I_{sp}^{i,0\,(\mathrm{II})}\right)\right]\approx\exp\left[-\lambda_s\omega_s\exp\left(-\frac{d^\alpha N_0\theta_s}{\sigma_s}\right)\right.
$$
$$
\left.\times\exp\left(-\lambda_p^*\pi\rho^2\exp\left(-\frac{D^\alpha N_0\theta_p}{\sigma_p}\right)\right)\right],
$$
(4.41)

where

$$
\omega_s=\pi\left\{\left(\frac{\sigma_s\theta_p}{\sigma_p}\right)^{\frac{2}{\alpha}}D^2\mathbb{E}_{h_{sp}^{j,0}}\left[(h_{sp}^{j,0})^{\frac{2}{\alpha}}\gamma\left(1-\frac{2}{\alpha},\theta_ph_{sp}^{j,0}\frac{\sigma_s}{\sigma_p}\left(\frac{D}{\rho}\right)^\alpha\right)\right]\right.
$$
$$
\left.-\frac{\sigma_sD^\alpha\theta_p\rho^{2-\alpha}}{\sigma_p+\sigma_sD^\alpha\theta_p\rho^{-\alpha}}\right\}
$$
(4.42)

and $\gamma(a,x)=\int_0^x\exp\left(-t\right)t^{a-1}dt$.

We substitute (4.39) and (4.41) into (4.38) and obtain the approximated PU success probability as

$$p_p^{'(\mathrm{II})} \approx \exp\left(-\frac{D^\alpha \theta_p N_0}{\sigma_p}\right) \exp\left[-\lambda_p^* C D^2 \left(\theta_p\right)^{\frac{2}{\alpha}} \exp\left(-\frac{D^\alpha \theta_p N_0}{\sigma_p}\right)\right]$$
$$\times \exp\left[-\lambda_s \omega_s \exp\left(-\frac{d^\alpha N_0 \theta_s}{\sigma_s}\right) \exp\left(-\lambda_p^* \pi \rho^2 \exp\left(-\frac{D^\alpha N_0 \theta_p}{\sigma_p}\right)\right)\right]. \quad (4.43)$$

Then, the approximation of the ASE of PU is

$$\bar{R}_p^{'(\mathrm{II})} = \lambda_p^* p_p^{'(\mathrm{II})} \log\left(1 + \theta_p\right). \quad (4.44)$$

Based on (4.11) and (4.44), the primary efficiency loss ratio is given by

$$\delta^{(\mathrm{II})} = \frac{\bar{R}_p - \bar{R}_p^{'(\mathrm{II})}}{\bar{R}_p}$$
$$\approx 1 - \exp\left[-\lambda_s \omega_s \exp\left(-\frac{d^\alpha \theta_s N_0}{\sigma_s}\right) \exp\left(-\lambda_p^* \pi \rho^2 \exp\left(-\frac{D^\alpha N_0 \theta_p}{\sigma_p}\right)\right)\right]. \quad (4.45)$$

Remark 4.3 According to (4.22) and (4.45), the primary efficiency loss ratio in Scheme II is equivalent to that of Scheme I for $\rho \to 0$. As ρ increases, the primary efficiency loss ratio in Scheme II is less than that of Scheme I since $\delta^{(\mathrm{II})}$ is a decreasing function of ρ. Larger ρ causes less degradation to the primary ASE, which provides better protection to the PUs.

4.6.2 SU Area Spectral Efficiency

Consider a typical secondary transmitter ST_0 and receiver SR_0 with SR_0 at the origin. ST_0 is active if the received $SINR_s$ from SR_0 is above θ_s while ST_0 is outside the active PERs. Given ST_0 is active, the SINR at SR_0 is given by

$$SINR_s^{(\mathrm{II})} = \frac{\sigma_s h_{ss}^{0,0} d^{-\alpha}}{\sum\limits_{x_j \in \Phi_p} I_{ps}^{j,0\ (\mathrm{II})} + \sum\limits_{y_i \in \Phi_s} I_{ss}^{i,0\ (\mathrm{II})} + N_0}, \quad (4.46)$$

where

$$I_{ps}^{j,0\ (\mathrm{II})} = \sigma_p h_{ps}^{j,0} |x_j|^{-\alpha} \mathbb{1}\left(h_{pp}^{j,j} \geq \frac{D^\alpha N_0 \theta_p}{\sigma_p}\right) \quad (4.47)$$

and

$$I_{ss}^{i,0\ (\mathrm{II})} = \sigma_s h_{ss}^{i,0} |y_i|^{-\alpha} \mathbb{1}\left(h_{ss}^{i,i} \geq \frac{d^\alpha N_0 \theta_s}{\sigma_s}\right) \mathbb{1}\ (ST_i \text{ is outside the active PERs}) \quad (4.48)$$

are the interference from PT_j and ST_i to SR_0, respectively. The secondary success probability is given by

$$p_s^{(II)} = \Pr\left(\mathrm{SINR}_s^{(II)} \geq \theta_p, h_{ss}^{0,0} \geq \frac{d^\alpha N_0 \theta_s}{\sigma_s}, ST_0 \text{ is outside the active PERs}\right).$$

(4.49)

Due to independence of events, we further have

$$p_s^{(II)} = \Pr\left(\mathrm{SINR}_s^{(II)} \geq \theta_p, h_{ss}^{0,0} \geq \frac{d^\alpha N_0 \theta_s}{\sigma_s}\right) \Pr\left(ST_0 \text{ is outside the active PERs}\right).$$

We assume the interference from the PTs and STs are independent,

$$p_s^{(II)} = \exp\left(-\frac{d^\alpha \theta_s N_0}{\sigma_s}\right) \mathbb{E}\left[\exp\left(-d^\alpha \theta_s \sum_{x_j \in \Phi_p} I_{ps}^{j,0\ (II)}\right)\right]$$

$$\times \mathbb{E}\left[\exp\left(-d^\alpha \theta_s \sum_{y_i \in \Phi_s} I_{ss}^{i,0\ (II)}\right)\right] \exp\left[-\lambda_p^* \pi \rho^2 \exp\left(-\frac{D^\alpha N_0 \theta_p}{\sigma_p}\right)\right]. \quad (4.50)$$

To further derive (4.50), we firstly discuss the interference from the active PTs to the SR_0. As discussed, the density of the active PTs is $\lambda_p^* \exp\left(-\frac{D^\alpha N_0 \theta_p}{\sigma_p}\right)$. Since an active ST_0 should stay at lease distance ρ from the active PRs, the minimum distance between the active PTs and active SR_0 is $\rho_0 = \max(0, \rho - D - d)$. Similar to (4.41), we also apply [8, (3.46)] to approximate the interference from the active PTs to SR_0. Since the active PTs are isotropic around ST_0 instead of SR_0, this approximation may include more interference than expected. Hence, the lower bound of the Laplace transform of the aggregate primary interference is given by

$$\mathbb{E}\left[\exp\left(-d^\alpha \theta_s \sum_{x_j \in \Phi_p} I_{ps}^{j,0\ (II)}\right)\right] \geq \exp\left[-\lambda_p^* \omega_p \exp\left(-\frac{D^\alpha \theta_p N_0}{\sigma_p}\right)\right], \quad (4.51)$$

where

$$\omega_p = \pi\left\{\left(\frac{\sigma_p \theta_s}{\sigma_s}\right)^{\frac{2}{\alpha}} d^2 \mathbb{E}_{h_{ps}^{j,0}}\left[(h_{ps}^{j,0})^{\frac{2}{\alpha}} \gamma\left(1 - \frac{2}{\alpha}, \theta_s h_{ps}^{j,0} \frac{\sigma_p}{\sigma_s}\left(\frac{d}{\rho_0}\right)^\alpha\right)\right]\right.$$

$$\left. -\frac{\sigma_p d^\alpha \theta_s \rho_0^{2-\alpha}}{\sigma_s + \sigma_p d^\alpha \theta_s \rho_0^{-\alpha}}\right\}.$$

Since $\rho_0 \neq 0$ in the above equation, we adopt $\rho_0 = \max\left(10^{-a}, \rho - D - d\right)$ instead, where a is a large integer.

In the next step, we derive the interference from other active STs to SR_0. As discussed, the active SUs follow Poisson hole process with density

$$\lambda_s \exp\left(-\frac{d^\alpha N_0 \theta_s}{\sigma_s}\right) \exp\left[-\lambda_p^* \pi \rho^2 \exp\left(-\frac{D^\alpha N_0 \theta_p}{\sigma_p}\right)\right]. \quad (4.52)$$

This could be well approximated by the Poisson cluster process [14], which how-ever, also does not lead to a closed-form Laplace transform. Therefore, we still approximate the active SUs as a homogeneous PPP with the same density as the Poisson hole process. Since the secondary nodes are more concentrated in Poisson hole process than in homogeneous PPP, the dispersion of the secondary nodes in the approximation may make the nearby interference smaller than expected. Hence, this approximation provides an upper bound to the Laplace transform of the aggregate secondary interference.

$$\mathbb{E}\left[\exp\left(-d^\alpha\theta_s\sum_{y_i\in\Phi_s} I_{ss}^{i,0\,\text{(II)}}\right)\right] \leq \exp\left[-\lambda_s Cd^2\,(\theta_s)^{\frac{2}{\alpha}}\exp\left(-\frac{d^\alpha N_0\theta_s}{\sigma_s}\right)\right.$$
$$\left. \times\exp\left(-\lambda_p^*\pi\rho^2\exp\left(-\frac{D^\alpha N_0\theta_p}{\sigma_p}\right)\right)\right]. \quad (4.53)$$

The lower bound can be calculated by considering all STs are active [14].

We substitute (4.51) and (4.53) into (4.50) and obtain the approximation of the secondary success probability, i.e.,

$$p_s^{\text{(II)}} \approx \exp\left(-\frac{d^\alpha\theta_s N_0}{\sigma_s}\right)\exp\left[-\lambda_p^*(\omega_p+\pi\rho^2)\exp\left(-\frac{D^\alpha\theta_p N_0}{\sigma_p}\right)\right]$$
$$\times\exp\left[-\lambda_s Cd^2\,(\theta_s)^{\frac{2}{\alpha}}\exp\left(-\frac{d^\alpha N_0\theta_s}{\sigma_s}\right)\exp\left(-\lambda_p^*\pi\rho^2\exp\left(-\frac{D^\alpha N_0\theta_p}{\sigma_p}\right)\right)\right].$$
$$(4.54)$$

Then, the approximation of the secondary ASE is given by

$$\bar{R}_s^{\text{(II)}} = \lambda_s p_s^{\text{(II)}}\log(1+\theta_s). \quad (4.55)$$

Remark 4.4 According to (4.26) and (4.54), the secondary success probability of Scheme II is equivalent to that of Scheme I for $\rho \to 0$. As ρ increases, the secondary success probability of Scheme II is less than that of Scheme I since $p_s^{\text{(II)}}$ is a decreasing function of ρ. As shown in *Remark 4.3*, larger ρ brings better protection to the PUs, which accommodates more concurrent secondary transmissions and may improve the secondary ASE. However, more concurrent secondary transmissions introduce more interference, which reduces the secondary success probability and restricts the secondary ASE. Therefore, a potential tradeoff exists between increasing the secondary success probability and reducing the primary efficiency loss ratio.

In Fig. 4.3, we show the analytical and simulation results of the primary and secondary success probabilities for Scheme I and Scheme II. The theoretical results match well with the simulation results for Scheme I and are close to the simulation results for Scheme II. Scheme II has higher primary success probability than Scheme I, which validates *Remark 4.3*. Moreover, Scheme II has lower the secondary success probability than Scheme I, which validates *Remark 4.4*.

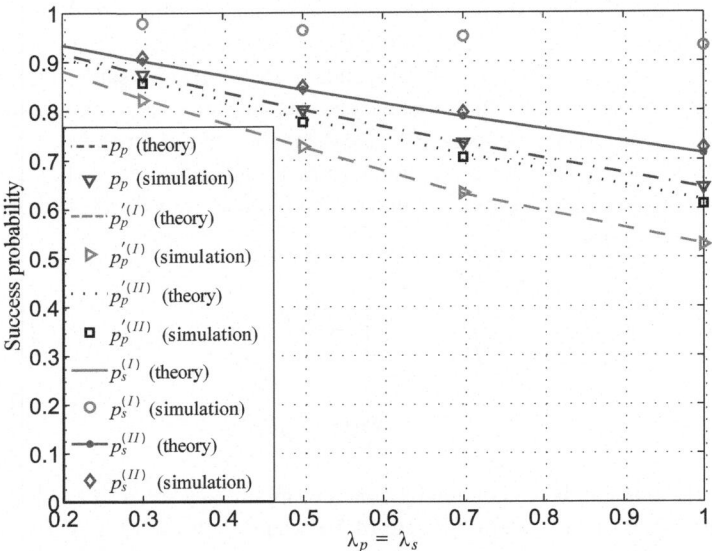

Fig. 4.3 Success probabilities ($D = 0.2$, $d = 0.05$, $\rho = 0.3$, $\alpha = 4$, $\theta_p = 5$, $\theta_s = 3$, $\sigma_p = 1$, $\sigma_s = 0.2$ and $N_0 = 10^{-3}$)

4.6.3   Optimal SU Density

Substituting (4.10), (4.45), (4.54) and (4.55) into $P4.2$ and solving it via KKT conditions, we obtain the optimal secondary density as follows.

(1) If $\varepsilon_s \leq \eta^{(\mathrm{II})}(r_{th})$,

$$\lambda_s^{*(\mathrm{II})} = \frac{\left[\frac{\left[\log\left(1-\varepsilon_p\right) + \frac{D^\alpha N_0 \theta_p}{\sigma_p} \right]\left(\omega_p + \pi\rho^2\right)}{C D^2 (\theta_p)^{\frac{2}{\alpha}}} - \log\left(1 - \varepsilon_s\right) - \frac{d^\alpha \theta_s N_0}{\sigma_s} \right]^+}{C d^2 (\theta_s)^{\frac{2}{\alpha}} \exp\left(-\frac{d^\alpha N_0 \theta_s}{\sigma_s}\right)}$$

$$\times \left[\frac{\exp\left(-\frac{D^\alpha N_0 \theta_p}{\sigma_p}\right)}{1 - \varepsilon_p} \right]^{\frac{\pi\rho^2}{C D^2 (\theta_p)^{\frac{2}{\alpha}}}} \tag{4.56}$$

$$\bar{R}_s^{(\mathrm{II})} = \frac{\left[\frac{\left[\log\left(1-\varepsilon_p\right) + \frac{D^\alpha N_0 \theta_p}{\sigma_p} \right]\left(\omega_p + \pi\rho^2\right)}{C D^2 (\theta_p)^{\frac{2}{\alpha}}} - \log\left(1 - \varepsilon_s\right) - \frac{d^\alpha \theta_s N_0}{\sigma_s} \right]^+}{C d^2 (\theta_s)^{\frac{2}{\alpha}} \exp\left(-\frac{d^\alpha N_0 \theta_s}{\sigma_s}\right)}$$

$$\times \left[\frac{\exp\left(-\frac{D^\alpha N_0 \theta_p}{\sigma_p}\right)}{1 - \varepsilon_p} \right]^{\frac{\pi\rho^2}{C D^2 (\theta_p)^{\frac{2}{\alpha}}}} (1 - \varepsilon_s) \log\left(1 + \theta_s\right), \tag{4.57}$$

(2) If $\varepsilon_s \geq \eta^{(\mathrm{II})}(r_{th})$,

$$\lambda_s^{*(\mathrm{II})} = \frac{-\log(1 - r_{th})}{\omega_s \left[\dfrac{1 - \varepsilon_p}{\exp\left(-\dfrac{D^{\alpha} N_0 \theta_p}{\sigma_p}\right)}\right]^{\frac{\pi \rho^2}{CD^2(\theta_p)^{\frac{2}{\alpha}}}} \exp\left(-\dfrac{d^{\alpha} \theta_s N_0}{\sigma_s}\right)} \qquad (4.58)$$

$$\bar{R}_s^{(\mathrm{II})} = -\frac{\log(1 - r_{th})}{\omega_s} \left[\frac{1 - \varepsilon_p}{\exp\left(-\dfrac{D^{\alpha} N_0 \theta_p}{\sigma_p}\right)}\right]^{\frac{\omega_p}{CD^2(\theta_p)^{\frac{2}{\alpha}}}} (1 - r_{th})^{\frac{Cd^2(\theta_s)^{\frac{2}{\alpha}}}{\omega_s}} \log(1 + \theta_s),$$

$$(4.59)$$

where

$$\eta^{(\mathrm{II})}(r_{th}) = 1 - \left[\frac{1 - \varepsilon_p}{\exp\left(-\dfrac{D^{\alpha} N_0 \theta_p}{\sigma_p}\right)}\right]^{\frac{\omega_p + \pi \rho^2}{CD^2(\theta_p)^{\frac{2}{\alpha}}}} (1 - r_{th})^{\frac{Cd^2(\theta_s)^{\frac{2}{\alpha}}}{\omega_s}} \exp\left(-\frac{d^{\alpha} \theta_s N_0}{\sigma_s}\right).$$

$$(4.60)$$

Remark 4.5 If $0 \leq \varepsilon_s \leq \kappa_s^{(\mathrm{II})}$, we have $\lambda_s^{*(\mathrm{II})} = 0$ since $1 - p_s > \varepsilon_s$ always holds, where

$$\kappa_s^{(\mathrm{II})} = 1 - \left[\frac{1 - \varepsilon_p}{\exp\left(-\dfrac{D^{\alpha} N_0 \theta_p}{\sigma_p}\right)}\right]^{\frac{\omega_p + \pi \rho^2}{CD^2(\theta_p)^{\frac{2}{\alpha}}}} \exp\left(-\frac{d^{\alpha} N_0 \theta_s}{\sigma_s}\right). \qquad (4.61)$$

The optimal secondary density is zero due to the tight SU outage constraint. If $\kappa_s^{(\mathrm{II})} < \varepsilon_s \leq \eta^{(\mathrm{II})}(r_{th})$, we have $\log(1 - \varepsilon_s) \approx -\varepsilon_s$ and $1 - \varepsilon_s \approx 1$. By further observation, $\lambda_s^{*(\mathrm{II})}$ and $\bar{R}_s^{(\mathrm{II})}$ increase almost linearly with ε_s. If $\varepsilon_s \geq \eta^{(\mathrm{II})}(r_{th})$, $\lambda_s^{*(\mathrm{II})}$ and $\bar{R}_s^{(\mathrm{II})}$ are constants with respect to ε_s.

Remark 4.6 For $\rho \to 0$, we have $\kappa_s^{(\mathrm{I})} \approx \kappa_s^{(\mathrm{II})}$. As ρ increases, we have $\kappa_s^{(\mathrm{II})}$ increases. According to *Remark 4.4*, the increase of ρ reduces the secondary success probability, which makes it harder to meet the tight secondary outage constraint ε_s. In this case, $\lambda_s^{(\mathrm{II})}$ and $\bar{R}_s^{(\mathrm{II})}$ remain zero for wider range of $\varepsilon_s \in (0, \kappa_s^{(\mathrm{II})}]$. Therefore, it is preferable to use relatively small PER radius for tight secondary outage constraint.

Remark 4.7 Moreover, $\eta^{(\mathrm{II})}(r_{th})$ is also an increasing function of ρ. The increase of ρ reduces $\delta^{(\mathrm{II})}$ and better protects the PU, according to *Remark 4.5*. There is more room for $\lambda_s^{*(\mathrm{II})}$ and $\bar{R}_s^{(\mathrm{II})}$ to increase with ε_s before the primary ELC is activated. Therefore, it is better to choose a relatively large PER radius for loose secondary outage constraint or tight primary ELC.

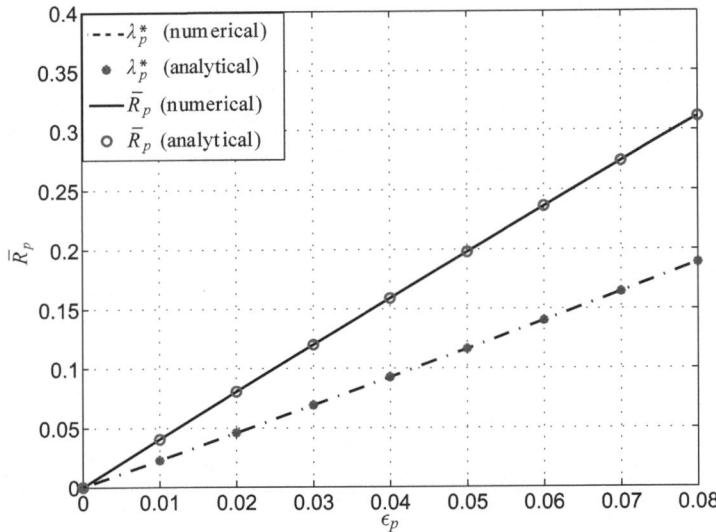

Fig. 4.4 Optimal density and maximized ASE of the PU ($D = 0.2$, $\alpha = 4$, $\theta_p = 5$, $\sigma_p = 1$ and $N_0 = 10^{-3}$)

4.7 Numerical Results

In Fig. 4.4, the numerical results of the optimal primary density λ_p^* and the maximized primary ASE $\bar{R}_p$ in $P4.1$ match well with the analytical results given in (4.10) and (4.11), respectively. For $\varepsilon_p \ll 1$, the optimal primary density and the maximized ASE increase almost linearly as the increase of the primary outage constraint. Intuitively, more primary transmissions are allowed when more outage can be tolerated by the system. The above phenomena validate *Remark 4.1*.

Figure 4.5 compares the numerical and analytical results of the optimal secondary density $\lambda_s^{*(I)}$ and the maximized secondary ASE $\bar{R}_s^{(I)}$ for Scheme I. For $\varepsilon_s \leq \eta^{(I)}(r_{th})$, the numerical results of $\lambda_s^{*(I)}$ and $\bar{R}_s^{(I)}$ match with the analytical results in (4.28) and (4.29), respectively. More specifically, for $0 \leq \varepsilon_s \leq \kappa_s^{(I)}$, we have $\lambda_s^{*(I)} = 0$ and $\bar{R}_s^{(I)} = 0$. In this case, the secondary outage probability cannot satisfy the tight outage constraint constraint. For $\kappa_s^{(I)} < \varepsilon_s \leq \eta^{(I)}(r_{th})$, $\lambda_s^{*(I)}$ and $\bar{R}_s^{(I)}$ increase almost linearly as the increase of ε_s. More concurrent active STs are allowed if the secondary outage constraint is less tight, which improves the secondary ASE. For $\varepsilon_s \geq \eta^{(I)}(r_{th})$, the curves of $\lambda_s^{*(I)}$ and $\bar{R}_s^{(I)}$ flatten out and match with (4.30) and (4.31), respectively. In this region, since the primary ELC dominates the problem, the secondary node density and ASE stops increasing even when more outage can be tolerated by the SU system. The above phenomena validate *Remark 4.2*.

In Fig. 4.6, the comparison is made between the numerical and analytical results of the optimal secondary density $\lambda_s^{*(II)}$ and the maximized ASE $\bar{R}_s^{(II)}$ for Scheme II. For $\varepsilon_s \leq \eta^{(II)}(r_{th})$, the numerical results of $\lambda_s^{*(II)}$ and $\bar{R}_s^{(II)}$ match with the analytical

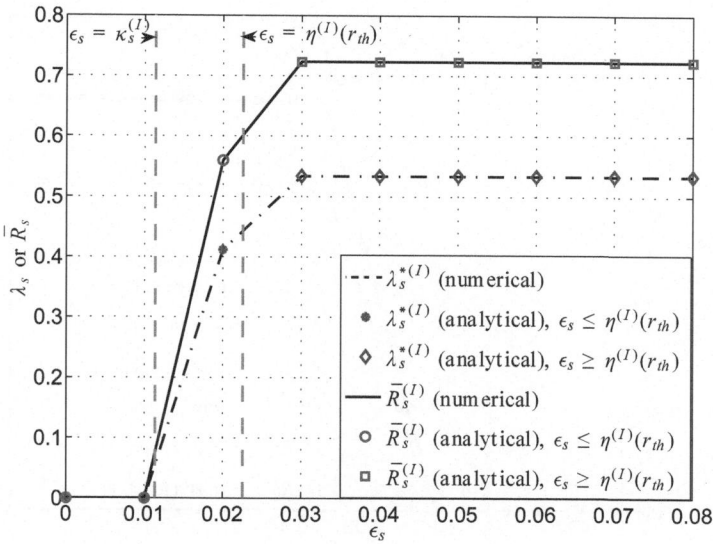

Fig. 4.5 Optimal density and maximized ASE of SU for Scheme I ($D = 0.2$, $d = 0.05$, $\alpha = 4$, $\theta_p = 5$, $\theta_s = 3$, $\sigma_p = 1$, $\sigma_s = 0.2$, $N_0 = 10^{-3}$, $\varepsilon_p = 0.1$ and $r_{th} = 0.1$)

results in (4.56) and (4.57), respectively. More specifically, for $0 \le \varepsilon_s \le \kappa_s^{(II)}$, we have $\lambda_s^* = 0$ and $\bar{R}_s = 0$. For $\kappa_s^{(II)} < \varepsilon_s \le \eta^{(II)}(r_{th})$, λ_s^* and $\bar{R}_s$ increase almost linearly as the increase of ε_s. For $\varepsilon_s \ge \eta^{(II)}(r_{th})$, the primary ELC dominates the problem. The curves of $\lambda_s^{*(II)}$ and $\bar{R}_s^{(II)}$ become flat and match (4.58) and (4.59), respectively. The above phenomena validate *Remark 4.5*.

Figure 4.7 compares Schemes I and II with different PER radii. Intuitively, Scheme I and Scheme II are equivalent if the PER radius is zero. However, $\rho \ne 0$ since it is on the denominator of ω_s. With ρ close to zero, i.e., $\rho = 10^{-5}$, the curve of $\bar{R}_s^{(II)}$ matches that of $\bar{R}_s^{(I)}$. Increasing ρ brings both benefit and loss to the SUs in Scheme II. On the one hand, increasing ρ makes $\lambda_s^{*(II)}$ and $\bar{R}_s^{(II)}$ remaining zero for larger range of ε_s, which validates that $\kappa_s^{(II)}$ is an increasing function of ρ in *Remark 4.6*. It is because larger ρ reduces the secondary success probability as shown in *Remark 4.4*. Thus, for a tight secondary outage constraint ε_s, it is preferable to have a relatively small PER radius. On the other hand, with relatively larger ρ, the primary ELC is activated at the point of larger ε_s, which allows $\bar{R}_s^{(II)}$ to increase more with ε_s. This validates that $\eta_s^{(II)}(r_{th})$ is also an increasing function of ρ in *Remark 4.7*. When ρ is relatively large, the PU is better protected, which reduces the chance for the primary efficiency loss $\delta^{(II)}$ to drop below r_{th} as shown in *Remark 4.3*. For tight primary ELC, it is more beneficial to use a relatively large PER radius.

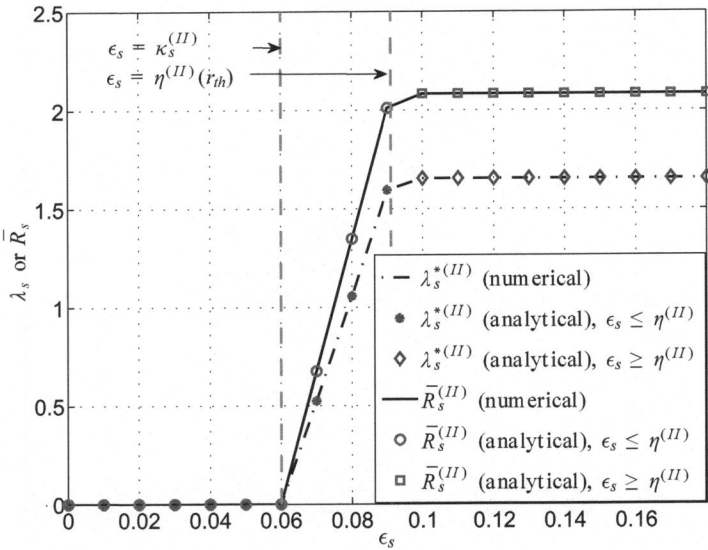

Fig. 4.6 Optimal density and maximized ASE of SU for Scheme II ($D = 0.2$, $d = 0.05$, $\alpha = 4$, $\theta_p = 5$, $\theta_s = 3$, $\sigma_p = 1$, $\sigma_s = 0.2$, $N_0 = 10^{-3}$, $\rho = 1.3D$, $\varepsilon_p = 0.1$ and $r_{th} = 0.1$)

4.8   Summary

This chapter discussed two opportunistic spectrum sharing schemes in the ad hoc network, without and with the PERs, respectively. The optimal density of the two schemes was derived analytically by maximizing the ASE of the SUs under the secondary outage constraint and primary ELC. For both schemes, the optimal node density and maximized ASE of the SUs increase almost linearly with the secondary outage constraint before they encounter the primary ELC. Scheme I with no PER is recommended if the SUs have tight secondary outage constraint or no primary location information. Scheme II with PER is preferred if the primary location and feedback information is available at the SUs and the primary ELC is tight.

Appendix A - Derivation of (4.6)

We rewrite p_p as

$$p_p = \Pr\left(\frac{\sigma_p h_{pp}^{0,0} D^{-\alpha}}{\sum\limits_{x_j \in \Phi_p} I_{pp}^{j,0} + N_0} \geq \theta_p, h_{pp}^{0,0} \geq \frac{D^\alpha N_0 \theta_p}{\sigma_p}\right)$$

$$= \Pr\left[h_{pp}^{0,0} \geq \frac{D^\alpha \theta_p}{\sigma_p}\left(\sum\limits_{x_j \in \Phi_p} I_{pp}^{j,0} + N_0\right)\right]$$

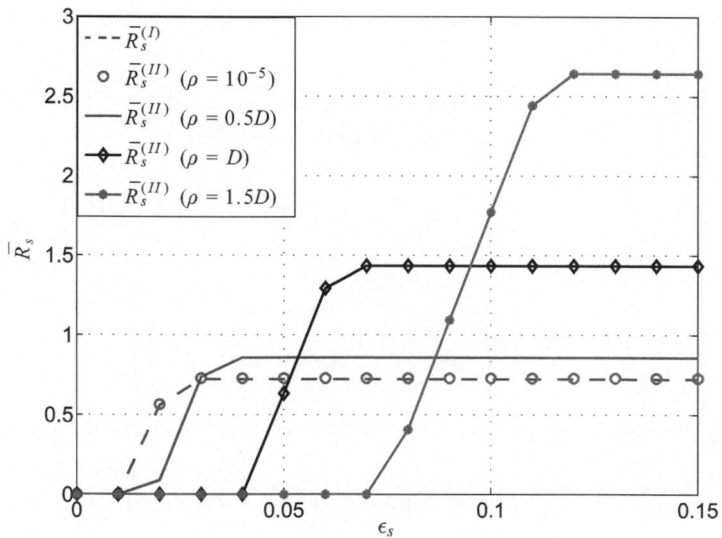

Fig. 4.7 Comparison between Scheme I and Scheme II ($D = 0.2$, $d = 0.05$, $\alpha = 4$, $\theta_p = 5$, $\theta_s = 3$, $\sigma_p = 1$, $\sigma_s = 0.2$, $N_0 = 10^{-3}$, $\varepsilon_p = 0.1$ and $r_{th} = 0.1$)

$$= \mathbb{E}\left[\exp\left(-\frac{D^\alpha \theta_p}{\sigma_p}\left(\sum_{x_j \in \Phi_p} I_{pp}^{j,0} + N_0\right)\right)\right]$$

$$= \beta \exp\left(-\frac{D^\alpha N_0 \theta_p}{\sigma_p}\right), \tag{4.62}$$

where $\beta = \mathbb{E}\left[\exp\left(-\frac{D^\alpha \theta_p}{\sigma_p}\sum_{x_j \in \Phi_p} I_{pp}^{j,0}\right)\right]$. Using the property of the exponential function and the independence of channel fading, we have

$$\beta = \mathbb{E}\left[\prod_{x_j \in \Phi_p}\exp\left(-\frac{D^\alpha \theta_p}{\sigma_p} I_{pp}^{j,0}\right)\right] = E\left[\prod_{x_j \in \Phi_p} E_{h_{pp}^{j,j}}\left[E_{h_{pp}^{j,0}}\left[\exp\left(-\frac{D^\alpha \theta_p}{\sigma_p} I_{pp}^{j,0}\right)\right]\right]\right] \tag{4.63}$$

where $I_{pp}^{j,0} = \sigma_p h_{pp}^{j,0}|x_j|^{-\alpha} \mathbb{1}\left(h_{pp}^{j,j} \geq \frac{D^\alpha N_0 \theta_p}{\sigma_p}\right)$. Let $s = D^\alpha \theta_p |x_j|^{-\alpha} \mathbb{1}\left(h_{pp}^{j,j} \geq \frac{D^\alpha N_0 \theta_p}{\sigma_p}\right)$. Since $h_{pp}^{j,0}$ follows exponential distribution with unit mean, the Laplace transform of $h_{pp}^{j,0}$ evaluated at s is

$$\mathcal{L}_{h_{pp}^{j,0}}(s) = \mathbb{E}_{h_{pp}^{j,0}}\left[\exp\left(-s h_{pp}^{j,0}\right)\right] = \frac{1}{1+s} = \frac{1}{1 + D^\alpha \theta_p |x_j|^{-\alpha} \mathbb{1}\left(h_{pp}^{j,j} \geq \frac{D^\alpha N_0 \theta_p}{\sigma_p}\right)}. \tag{4.64}$$

Substituting (4.64) into (4.63) yields

$$
\begin{aligned}
\beta &= \mathbb{E}\left[\prod_{x_j \in \Phi_p} \mathbb{E}_{h_{pp}^{j,j}}\left[\frac{1}{1 + D^\alpha \theta_p |x_j|^{-\alpha} \mathbb{1}\left(h_{pp}^{j,j} \geq \frac{D^\alpha N_0 \theta_p}{\sigma_p}\right)}\right]\right] \\
&= \mathbb{E}\left[\prod_{x_j \in \Phi_p}\left[\int_0^{\frac{D^\alpha N_0 \theta_p}{\sigma_p}} \exp\left(-h_{pp}^{j,j}\right) dh_{pp}^{j,j} + \int_{\frac{D^\alpha N_0 \theta_p}{\sigma_p}}^\infty \frac{\exp\left(-h_{pp}^{j,j}\right)}{1 + D^\alpha \theta_p |x_j|^{-\alpha}} dh_{pp}^{j,j}\right]\right] \\
&= \mathbb{E}\left[\prod_{x_j \in \Phi_p}\left[1 - \exp\left(-\frac{D^\alpha N_0 \theta_p}{\sigma_p}\right) + \frac{\exp\left(-\frac{D^\alpha N_0 \theta_p}{\sigma_p}\right)}{1 + D^\alpha \theta_p |x_j|^{-\alpha}}\right]\right].
\end{aligned}
\tag{4.65}
$$

Applying the probability generating functional (PGFL) [4][Proposition 2.12]

$$
\mathbb{E}\prod_{x_j \in \Phi_p} f(x) = \exp\left(-\lambda_p \int_{\mathbb{R}^2} [1 - f(x)] dx\right),
\tag{4.66}
$$

we have

$$
\begin{aligned}
\beta &= \exp\left[-\lambda_p \int_{\mathbb{R}^2}\left[1 - \left(1 - \exp\left(-\frac{D^\alpha N_0 \theta_p}{\sigma_p}\right) + \frac{\exp\left(-\frac{D^\alpha N_0 \theta_p}{\sigma_p}\right)}{1 + D^\alpha \theta_p |x_j|^{-\alpha}}\right)\right] dx\right] \\
&= \exp\left[-\lambda_p \exp\left(-\frac{D^\alpha N_0 \theta_p}{\sigma_p}\right) \int_{\mathbb{R}^2}\left(1 - \frac{1}{1 + D^\alpha \theta_p |x_j|^{-\alpha}}\right) dx\right].
\end{aligned}
\tag{4.67}
$$

Carry out a transformation to spherical polar coordinates and denote radial coordinate and angular coordinate by v and ψ, respectively, (4.67) is rewritten as

$$
\begin{aligned}
\beta &= \exp\left[-\lambda_p \exp\left(-\frac{D^\alpha N_0 \theta_p}{\sigma_p}\right) \int_0^{2\pi} \int_0^\infty \frac{D^\alpha \theta_p v^{1-\alpha}}{1 + D^\alpha \theta_p v^{-\alpha}} dv d\psi\right] \\
&= \exp\left[-\lambda_p \exp\left(-\frac{D^\alpha N_0 \theta_p}{\sigma_p}\right) 2\pi \int_0^\infty \frac{D^\alpha \theta_p v^{1-\alpha}}{1 + D^\alpha \theta_p v^{-\alpha}} dv\right] \\
&= \exp\left[-\lambda_p \exp\left(-\frac{D^\alpha N_0 \theta_p}{\sigma_p}\right) \frac{2\pi^2 D^2 (\theta_p)^{\frac{2}{\alpha}}}{\alpha \sin\left(\frac{2\pi}{\alpha}\right)}\right].
\end{aligned}
\tag{4.68}
$$

Substituting (4.68) into (4.62) yields

$$
p_p = \exp\left[-\lambda_p \exp\left(-\frac{D^\alpha N_0 \theta_p}{\sigma_p}\right) \frac{2\pi^2 D^2 (\theta_p)^{\frac{2}{\alpha}}}{\alpha \sin\left(\frac{2\pi}{\alpha}\right)}\right] \exp\left(-\frac{D^\alpha N_0 \theta_p}{\sigma_p}\right).
\tag{4.69}
$$

References

1. Z. Wang and W. Zhang, "Spectrum sharing with limited feedback in Poisson cognitive network," in Proc. *IEEE Int. Conf. on Commun. (ICC 2014)*, Sydney, Australia, pp. 1441–1446, June 10–14, 2014.
2. Z. Wang and W. Zhang, "Opportunistic spectrum sharing with limited feedback in Poisson cognitive radio networks," *IEEE Trans. Wireless Commun.*, vol. 13, no. 12, pp. 7098–7109, Dec. 2014.
3. D. Stoyan, W. Kendall, and J. Mecke, *Stochastic Geometry and its Applications, 2nd Edition,* Wiley, 1995.
4. S. P. Weber, X. Yang, J. G. Andrews, and G. de Veciana, "Transmission capacity of wireless ad hoc networks with outage constraints," *IEEE Trans. Inf. Theory*, vol. 51, pp. 4091–4102, Dec. 2005.
5. S. P. Weber, J. G. Andrews, and N. Jindal, "An overview of transmission capacity of wireless networks," *IEEE Trans. Wireless Commun.*, vol. 58, pp. 3593–3604, Dec. 2010.
6. S. Weber, J. G. Andrews, and N. Jindal, "The effect of fading, channel inversion, and threshold scheduling on ad hoc networks," *IEEE Trans. Inf. Theory*, vol. 53, pp. 4127–4149, Nov. 2007.
7. M. Haenggi, J. G. Andrews, F. Baccelli, O. Dousse, and M. Franceschetti, "Stochastic geometry and random graphs for the analysis and design of wireless networks," *IEEE J. Sel. Areas Commun.*, vol. 27, pp. 1029–1046, Sep. 2009.
8. M. Haenggi and R. K. Ganti, *Interference in Large Wireless Networks,* Now Publishers, 2008.
9. S. Weber and J. G. Andrews, *Transmission Capacity of Wireless Networks,* Now Publishers, 2012.
10. H. ElSawy, E. Hossain, and M. Haenggi, "Stochastic geometry for modeling, analysis, and design of multi-tier and cognitive cellular wireless networks: A survey," *IEEE Commun. Surveys & Tutorials,* vol. 15, pp. 996–1019, Third Quater, 2013.
11. M. Vu, N. Devroye, and V. Tarokh, "On the primary exclusive region of cognitive networks," *IEEE Trans. Wireless Commun.*, vol. 8, pp. 3380–3385, Jul. 2009.
12. Y. Kim and G. de Veciana "Joint network capacity region for cognitive networks heterogeneous environments and RF-environment awareness," *IEEE J. Sel. Areas Commun.*, vol. 29, pp. 407–420, Feb. 2011.
13. X. Song, C. Yin, D. Liu, and R. Zhang, "Spatial opportunity in cognitive radio networks with threshold-based opportunistic spectrum access," in *Proc. IEEE Int. Conf. Commun.*, Budapest, Hungary, Jun. 9–15, 2013.
14. C. Lee and M. Haenggi, "Interference and outage Poisson cognitive networks," *IEEE Trans. Wireless Commun.*, vol. 11, no. 4, pp. 1392–1401, Apr. 2012.
15. L. Wang and V. Fodor, "On the gain of primary exclusion region and vertical cooperation in spectrum sharing wireless networks," *IEEE Trans. Veh. Technol.*, vol. 61, no. 8, pp. 3746–3758, Oct. 2012.
16. S.-W. Jeon, N. Devroye, M. Vu, S.-Y. Chung, and V. Tarokh, "Cognitive networks achieve throughput scaling of a homogeneous network," *IEEE Trans. Inf. Theory*, vol. 57, pp. 5103–5115, Nov. 2011.
17. L. Gao, R. Zhang, C. Yin, and S. Cui "Throughput and delay scaling in supportive two-tier networks," *IEEE J. Sel. Areas Commun.*, vol. 30, pp. 415–424, Feb. 2012.
18. R. Cai, W. Zhang, and P. C. Ching, "Spectrum sharing between random geometric networks," in Proc. *IEEE Int. Conf. Acoust., Speech and Signal Process. (ICASSP 2012)*, Kyoto, Japan, Mar. 25–30, 2012.
19. C. Yin, C. Chen, T. Liu, and S. Cui, "Generalized results of transmission capacities for overlaid wireless networks," in Proc. *IEEE Int. Symp. Inf. Theory*, Seoul, Korea, Jun. 28–Jul. 3, 2009.
20. F. Baccelli, B. Blaszczyszyn, and P. Muhlethaler, "Stochastic analysis of spatial and opportunistic Aloha," *IEEE J. Sel. Areas Commun.*, vol. 27, no. 7, pp. 1105–1119, Sep. 2009.

21. M. Fitch, M. Nekovee, S. Kawade, K. Briggs, and R. MacKenzie, "Wireless service provision in TV white space with cognitive radio technology: A telecom operator's perspective and experience," *IEEE Commun. Mag.*, vol. 49, pp. 64–73, Mar. 2011.
22. S. Boyd and L. Vandenberghe, *Convex Optimization,* Cambridge, U.K. Cambridge Univ. Press, 2004.

Chapter 5
Conclusions

Abstract In this chapter, we summarize main results and conclude the brief.

Charles Darwin said: "*It is not the strongest of the species that survives, nor the most intelligent that survives. It is the one that is most adaptable to change*". Cognitive radio improves the spectrum efficiency by allowing Secondary user (SU) to share the underutilized licensed spectrum of primary user (PU). For spectrum sharing, the key is to learn and adapt to the ever-changing spectrum environment. SU shares the spectrum with PU by detecting and utilizing the spectrum holes in temporal, spatial, angle and code domains. In practice, it is difficult for SU to obtain the spectrum holes information without the cooperation of PU. Limited feedback informs the transmitters of the instantaneous spectrum environment and enables opportunistic spectrum sharing. If PU is a rate adaptive or error control system, primary receiver (PR) sends some limited channel feedback to primary transmitter (PT). By eavesdropping the primary feedback, SU is able to know some partial information of the spectrum holes and adapts it transmission accordingly. If secondary receiver (SR) also sends some limited channel feedback to secondary transmitter (ST), SU can be more adaptive to the spectral environment, which enhances the secondary performance and improves the spectral utilization efficiency.

In this brief, we explored temporal and spatial spectrum holes by feedback based opportunistic spectrum sharing schemes in the point-to-point, broadcast scheduling and ad hoc cognitive networks. For these schemes, we presented the optimal resource allocation that maximizes the performance of the SU while maintaining an appropriate quality of service of the PU. In the point-to-point and broadcast scheduling networks, we derived the optimal quantization thresholds and power allocation that maximize the secondary throughput while not causing harmful degradation of the primary throughput. The secondary throughput was shown to: (1) increase with more secondary feedback bits and more side information of the interference from the PT; (2) scale double logarithmically as the increase of the number of the SRs. In the ad hoc network, with the aid of stochastic geometry, we derived the optimal secondary node density that maximizes the secondary throughput while satisfying the reliability targets of both the PU and SU. The primary location information is shown to be beneficial to the secondary throughput if the PU has a high reliability requirement.

© The Author(s) 2015
Z. Wang, W. Zhang, *Opportunistic Spectrum Sharing in Cognitive Radio Networks*,
SpringerBriefs in Electrical and Computer Engineering, DOI 10.1007/978-3-319-15542-5_5

In the future work, more precise feedback, more adaptive transmit power, general fading scenarios are to be addressed in more random network topologies. Furthermore, apart from exploiting temporal and spatial spectrum holes with limited feedback, other domains such as angle and code domains are still left open.